THE EXTREMITIES
Muscles and M

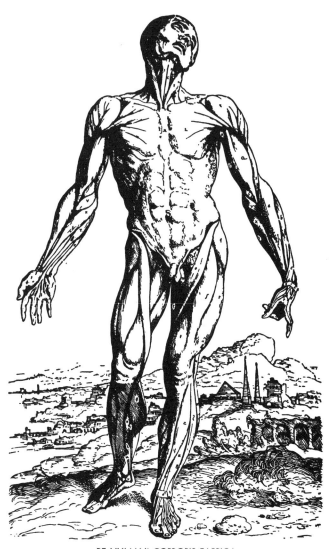

DE HUMANI CORPORIS FABRICA
ANDREAE VESALII
1568

THE EXTREMITIES
Muscles and Motor Points

JOHN H. WARFEL, Ph.D.

Associate Professor of Anatomy, Emeritus,
State University of New York at Buffalo
School of Medicine, Buffalo, New York

Sixth Edition

110 Illustrations

LEA & FEBIGER

Philadelphia and London

1993

Lea & Febiger
Box 3024
200 Chester Field Parkway
Malvern, Pennsylvania   19355-9725
U.S.A.
(215) 251-2230

Executive Editor: George H. Mundorff
Manuscript Editor: Jessica Howie Martin
Production Manager: Robert N. Spahr

First Edition, 1945
Reprinted
    June, 1946
    April, 1948
    September, 1950
    January, 1952
    October, 1953
    November, 1954
    September, 1956
    February, 1958

Second Edition, 1960
Reprinted
    August, 1963

Third Edition, 1967
Reprinted
    March, 1970

Fourth Edition, 1974
Reprinted
    April, 1976
    May, 1978
    January, 1981
    November, 1982

Fifth Edition, 1985
Reprinted
    January, 1987
    May, 1989

Sixth Edition, 1993

Library of Congress Cataloging in Publication Data

Warfel, John H. 1916–
    The extremities : muscles and motor points / John H. Warfel. — 6th ed.
        p.   cm.
    Includes bibliographical references and index.
    ISBN 0-8121-1582-1
    1. Extremities (Anatomy) 2. Muscles—Innervation. 3. Myoneural
junction. I. Title.
    [DNLM: 1. Extremities—anatomy & histology—atlases. 2. Motor
Endplate—anatomy & histology—atlases. 3. Muscles—anatomy &
histology—atlases. WE 17 W275e]
    QM165.W27   1993
    611'.738—dc20
    DNLM/DLC
    for Library of Congress   92–17503
                                                                   CIP

PRINTED IN THE UNITED STATES OF AMERICA

Print number: 5   4   3   2   1

Reprints of chapters may be purchased from Lea & Febiger in quantities of 100 or more.
Contact Sally Grande in the Sales Department.

PREFACE

In this sixth edition, as in the sixth edition of the companion volume, *The Head, Neck, and Trunk*, the references to the 30th American Edition of Gray's Anatomy (Philadelphia, Lea & Febiger, 1985) remain unchanged. The references to Grant's Atlas of Anatomy (Baltimore, The Williams & Wilkins Co., 1991) have been corrected to the new 9th edition.

This edition of The Extremities has been enhanced by the inclusion of references to the Atlas of Human Anatomy by Frank H. Netter, M.D. (Summit, New Jersey, Ciba-Geigy Corporation, 4th printing, 1991). Terminology throughout has been verified against the sixth edition of the Nomina Anatomica (New York, Churchill Livingstone, 1989).

Diagrammatic representation and condensed description cannot do full justice to the complex relations involved. The illustrations do not attempt to show all details of attachments, nerves, and arteries, since the object is to emphasize the major termini of muscles and the chief arteries and nerves related to them. Likewise, the legends stress the primary functions, which imply movement at the insertions. The student must realize, however, that when the insertions are fixed, muscles produce movement at their origins as well. In the lower extremity, this occurs almost as frequently as primary action.

Motor points were tested on normal subjects. Since the points vary among individuals, diagrams can give only the approximate location of greatest muscular response. Motor points are not included for muscles that do not show a clear-cut response to electrical stimulus.

Buffalo, New York John H. Warfel

CONTENTS

6. DORSAL MUSCLES OF THE FOREARM

Superficial Group

Deep Group

7. MUSCLES OF THE HAND

8. MUSCLES OF THE LOWER EXTREMITY, ILIAC REGION

9. ANTERIOR MUSCLES OF THE THIGH

15. LATERAL MUSCLES OF THE LEG

16. MUSCLES OF THE FOOT

CHARTS

1. MUSCLES CONNECTING THE UPPER EXTREMITY TO THE VERTEBRAL COLUMN

    Trapezius
    Latissimus dorsi
    Rhomboideus major
    Rhomboideus minor
    Levator scapulae

TRAPEZIUS

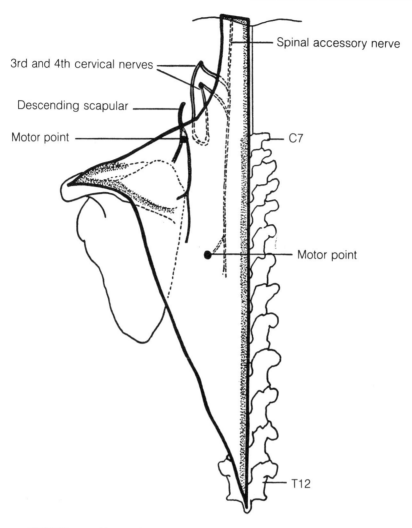

Spinal accessory nerve

3rd and 4th cervical nerves

Descending scapular

Motor point

C7

Motor point

T12

ORIGIN: External occipital protuberance, superior nuchal line, nuchal ligament from spines of seventh cervical and all thoracic vertebrae
INSERTION: Lateral third of clavicle, spine of scapula, acromion
FUNCTION: Adducts scapula, tilts chin, draws back acromion, rotates scapula
NERVE: Spinal Accessory, 3d and 4th cervical
ARTERY: Descending scapular (transverse cervical)

References

	GRAY	GRANT'S ATLAS	NETTER
Muscle	513	4-47, 6-32, 8-4	22, 160, 178
Nerve	513, 1189, 1199, 1200, 1205	4-47, 6-32, 6-33, 8-4, 9-16	112, 121, 163
Artery	704, 706	6-33	27, 163, 404

LATISSIMUS DORSI

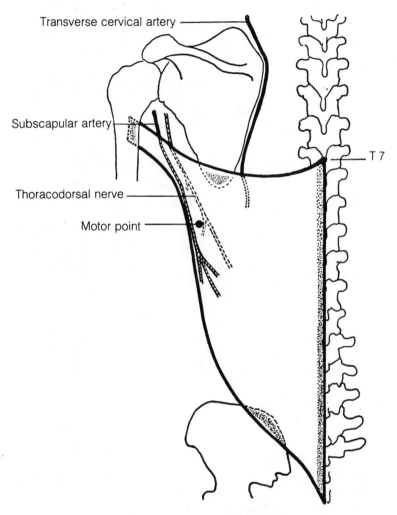

Transverse cervical artery

Subscapular artery

Thoracodorsal nerve

Motor point

T 7

ORIGIN: Spines of lower 6 thoracic vertebrae, lumbodorsal fascia, crest of ilium, muscular slips from lower 3 or 4 ribs

INSERTION: Floor of bicipital groove of humerus

FUNCTION: Adducts, extends, and medially rotates humerus

NERVE: Thoracodorsal

ARTERY: Descending scapular (transverse cervical), subscapular

References

	GRAY	GRANT'S ATLAS	NETTER
Muscle	513	4-47	160, 178, 401
Nerve	515, 1205, 1207, 1210	6-17, 6-25, 6-26	401, 404, 405
Artery	706, 714	6–25	Not shown

RHOMBOIDEUS MAJOR

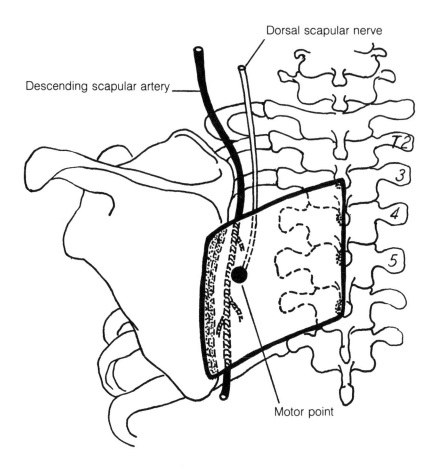

Dorsal scapular nerve

Descending scapular artery

T2

3

4

5

Motor point

ORIGIN: Spine of 2d, 3d, 4th and 5th thoracic vertebrae
INSERTION: Medial border of scapula, between spine and inferior angle
FUNCTION: Adducts and rotates scapula
NERVE: Dorsal scapular
ARTERY: Descending scapular (transverse cervical)

References

	GRAY	GRANT'S ATLAS	NETTER
Muscle	515	4-47, 4-48	160, 399
Nerve	516, 1205, 1207	8-4B, 8-4C	404, 405, 450
Artery	706	8-4B, 8-4C	Not shown

RHOMBOIDEUS MINOR

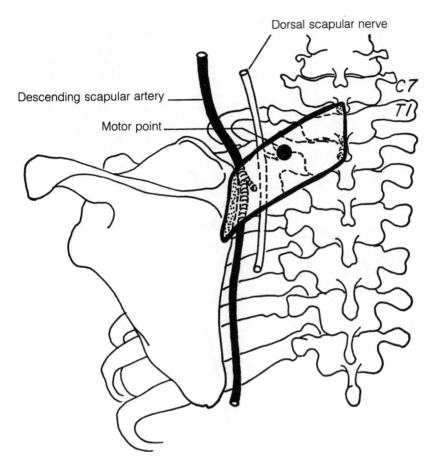

Dorsal scapular nerve

Descending scapular artery

Motor point

C7

T1

ORIGIN: Ligamentum nuchae, spine of 7th cervical and 1st thoracic
            vertebrae
INSERTION; Root of scapular spine
FUNCTION: Adducts and rotates scapula
NERVE: Dorsal scapular
ARTERY: Descending scapular (transverse cervical)

References

	GRAY	GRANT'S ATLAS	NETTER
Muscle	515	4-47, 4-48	160, 399
Nerve	516, 1205, 1207	8-4B, 8-4C	404, 405, 450
Artery	706	8-4B, 8-4C	Not shown

LEVATOR SCAPULAE

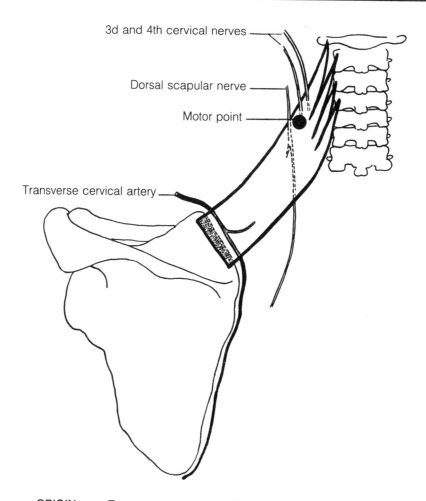

3d and 4th cervical nerves

Dorsal scapular nerve

Motor point

Transverse cervical artery

ORIGIN: Transverse process of atlas, axis, 3d and 4th cervical vertebrae

INSERTION: Vertebral border of scapula between superior (medial) angle and root of spine

FUNCTION: Raises scapula or inclines neck to corresponding side if scapula is fixed

NERVE: Dorsal scapular, 3d and 4th cervical

ARTERY: Descending scapular (transverse cervical)

References

	GRAY	GRANT'S ATLAS	NETTER
Muscle	516	4-47, 4-48	27, 160
Nerve	516, 1200, 1205, 1207	8-4B, 8-4C	404, 405, 450
Artery	706	8-4C, 8-4D	Not shown

2. MUSCLES CONNECTING UPPER EXTREMITY TO THE ANTERIOR AND LATERAL THORACIC WALLS

Pectoralis major
Pectoralis minor
Subclavius
Seratus anterior

PECTORALIS MAJOR

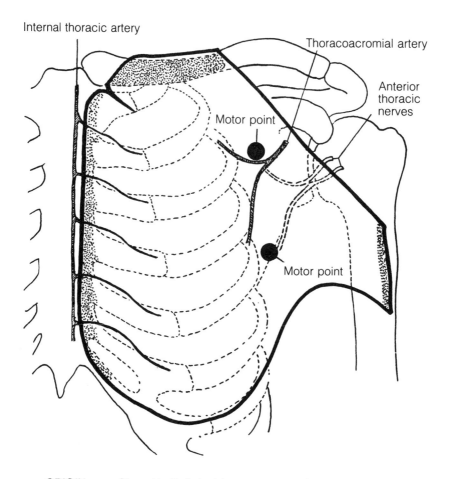

Internal thoracic artery

Thoracoacromial artery

Anterior thoracic nerves

Motor point

Motor point

ORIGIN: Sternal half of clavicle, sternum to 7th rib, cartilages of true ribs, aponeurosis of external oblique muscle

INSERTION: Lateral lip of bicipital groove of humerus

FUNCTION: Adducts arm, draws it forward, rotates it medially

NERVE: Medial and lateral anterior thoracic (medial and lateral pectoral)

ARTERY: Pectoral branch of thoracoacromial, perforating branches of internal thoracic

References

	GRAY	GRANT'S ATLAS	NETTER
Muscle	518	1-2, 6-13, 6-14	174, 232, 399
Nerve	519, 1205, 1207, 1209	6-14, 6-22	404, 405
Artery	707, 712, 713	6-14, 6-22	174, 404

PECTORALIS MINOR

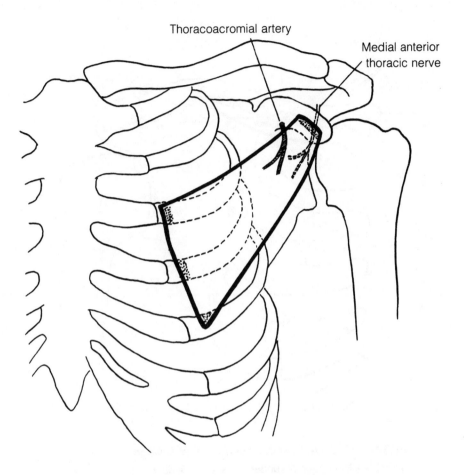

Thoracoacromial artery

Medial anterior thoracic nerve

ORIGIN: Outer surface of upper margin of 3d, 4th, and 5th rib
INSERTION: Coracoid process of scapula
FUNCTION: Lowers lateral angle of scapula, pulls shoulder forward
NERVE: Medial anterior thoracic (medial pectoral)
ARTERY: Thoracoacromial and intercostal branches of internal thoracic, lateral thoracic

References

	GRAY	GRANT'S ATLAS	NETTER
Muscle	520	1-15	175, 403, 408
Nerve	520, 1205, 1207, 1209	6-22	404, 405
Artery	707, 712, 713	6-22	175, 403

SUBCLAVIUS

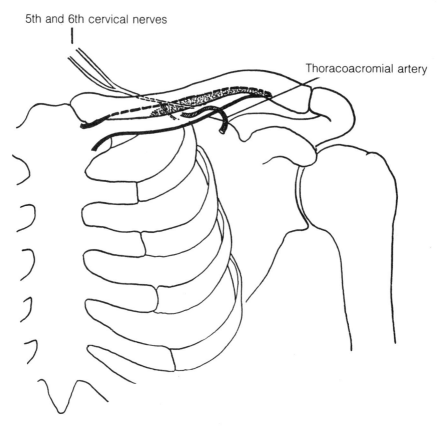

5th and 6th cervical nerves

Thoracoacromial artery

ORIGIN: Upper border of 1st rib and its cartilage
INSERTION: Groove on under surface of clavicle
FUNCTION: Draws clavicle down and forward
NERVE: 5th and 6th cervical (nerve to subclavius)
ARTERY: Clavicular branch of thoracoacromial

References

	GRAY	GRANT'S ATLAS	NETTER
Muscle	521	6-22, 8-4	175, 219, 403
Nerve	521, 1205, 1207, 1209	8-4D	405
Artery	712	6-14	404

SERRATUS ANTERIOR

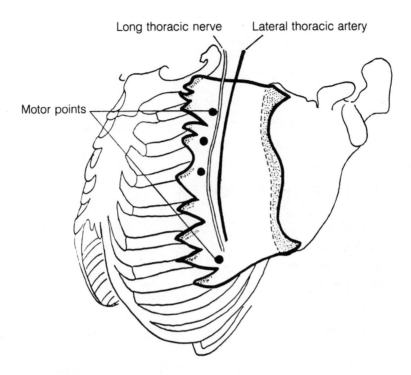

Long thoracic nerve   Lateral thoracic artery

Motor points

ORIGIN: Outer surface of upper 8 or 9 ribs
INSERTION: Costal surface of vertebral border of scapula
FUNCTION: Abducts scapula; raises ribs when scapula is fixed
NERVE: Long thoracic
ARTERY: Lateral thoracic

References

	GRAY	GRANT'S ATLAS	NETTER
Muscle	521	4-48, 6-28	177, 232
Nerve	521, 1205, 1207	2-7, 6-13, 6-26, 6-28	175, 404, 405
Artery	713	6-22	404

3. MUSCLES OF THE SHOULDER

Deltoideus
Subscapularis
Supraspinatus
Infraspinatus
Teres minor
Teres major

DELTOIDEUS

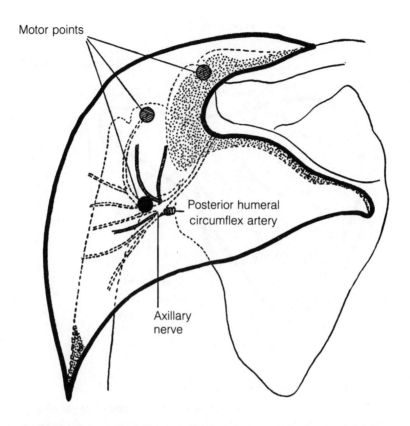

Motor points

Posterior humeral
circumflex artery

Axillary
nerve

ORIGIN: Lateral third of clavicle, upper surface of acromion, spine of scapula
INSERTION: Deltoid tuberosity of humerus
FUNCTION: Abducts arm
NERVE: Anterior and posterior branches of axillary (circumflex)
ARTERY: Posterior humeral circumflex, deltoid branch of thoracoacromial

References

	GRAY	GRANT'S ATLAS	NETTER
Muscle	522	6-14, 6-39	160, 163, 399
Nerve	522, 1205, 1207, 1210	6-27, 6-40	405, 407, 450
Artery	712, 714	6-40	401, 404

SUBSCAPULARIS

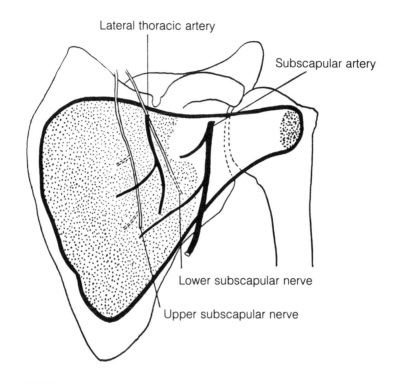

Lateral thoracic artery

Subscapular artery

Lower subscapular nerve

Upper subscapular nerve

ORIGIN: Subscapular fossa
INSERTION: Lesser tuberosity of humerus and capsule of shoulder joint
FUNCTION: Rotates humerus medially, draws it forward and down when arm is raised
NERVE: Upper and lower subscapular
ARTERY: Lateral thoracic, subscapular

References

	GRAY	GRANT'S ATLAS	NETTER
Muscle	522	6-26, 6-52	177, 400
Nerve	523, 1205, 1207, 1209	6-17, 6-26	401, 405
Artery	713, 714	6-22, 6-25	404

SUPRASPINATUS

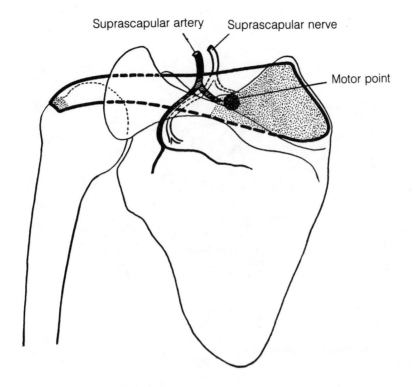

Suprascapular artery   Suprascapular nerve

Motor point

ORIGIN: Supraspinous fossa of scapula
INSERTION: Superior facet of greater tuberosity of humerus; capsule of shoul-
              der joint
FUNCTION: Assists deltoid in abducting arm, fixes head of humerus in glenoid
              cavity; rotates head of humerus laterally
NERVE: Suprascapular
ARTERY: Suprascapular (transverse scapular)

References

	GRAY	GRANT'S ATLAS	NETTER
Muscle	523	6-35, 6-52	160, 400
Nerve	523, 1205, 1207, 1209	6-33, 6-40	401, 405, 450
Artery	703	6-33	402

INFRASPINATUS

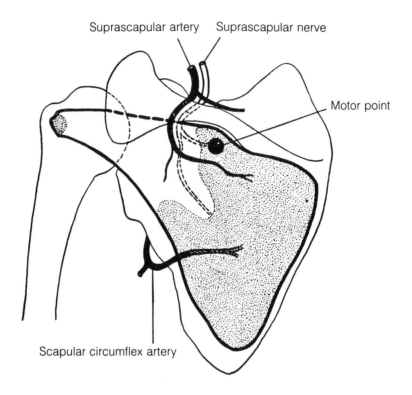

Suprascapular artery   Suprascapular nerve

Motor point

Scapular circumflex artery

ORIGIN: Infraspinous fossa of scapula
INSERTION: Middle facet of greater tuberosity of humerus; capsule of shoulder joint
FUNCTION: Rotates head of humerus laterally with teres minor
NERVE: Suprascapular
ARTERY: Suprascapular (transverse scapular); scapular circumflex

References

	GRAY	GRANT'S ATLAS	NETTER
Muscle	523	6-40, 6-52	160, 400, 450
Nerve	524, 1205, 1207, 1209	6-33, 6-40	401, 405, 450
Artery	703, 713	6-40	402

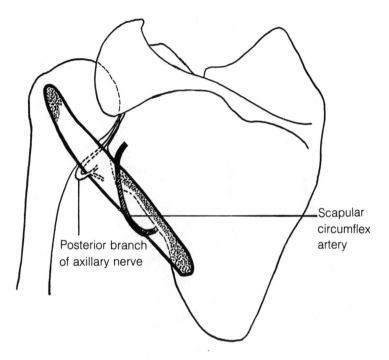

Scapular circumflex artery

Posterior branch of axillary nerve

ORIGIN: Dorsal surface of axillary border of scapula

INSERTION: Lowest facet of greater tuberosity of humerus; capsule of shoulder joint

FUNCTION: Adducts and rotates head of humerus laterally and draws humerus toward glenoid fossa

NERVE: Posterior branch of axillary (circumflex)

ARTERY: Scapular circumflex

References

	GRAY	GRANT'S ATLAS	NETTER
Muscle	524	6-40, 6-52	160, 400, 401
Nerve	524, 1205, 1207, 1210	6-27, 6-40	401, 405
Artery	713	6-40	402

TERES MAJOR

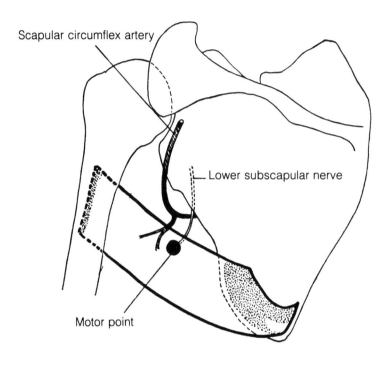

Scapular circumflex artery

Lower subscapular nerve

Motor point

ORIGIN: Dorsal surface of inferior angle of scapula
INSERTION: Medial lip of bicipital groove of humerus
FUNCTION: Adducts and medially rotates humerus and draws it back
NERVE: Lower subscapular
ARTERY: Scapular circumflex

References

	GRAY	GRANT'S ATLAS	NETTER
Muscle	524	6-26, 6-40	160, 178, 401
Nerve	524, 1205, 1207, 1209	6-17, 6-26	404, 405, 450
Artery	713	6-40	402

4. MUSCLES OF THE ARM

Coracobrachialis
Biceps brachii
Brachialis
Triceps brachii

CORACOBRACHIALIS

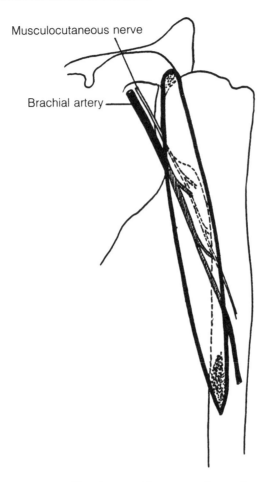

Musculocutaneous nerve

Brachial artery

ORIGIN: Tip of coracoid process of scapula
INSERTION: Middle of medial border of humerus
FUNCTION: Flexion and adduction of arm
NERVE: Musculocutaneous
ARTERY: Muscular branches of brachial

References

	GRAY	GRANT'S ATLAS	NETTER
Muscle	526	6-22, 6-26	401, 406
Nerve	526, 1205, 1207, 1212	6-26, 6-27, 6-43	404, 405, 408, 447
Artery	719	6-43	408

BICEPS BRACHII

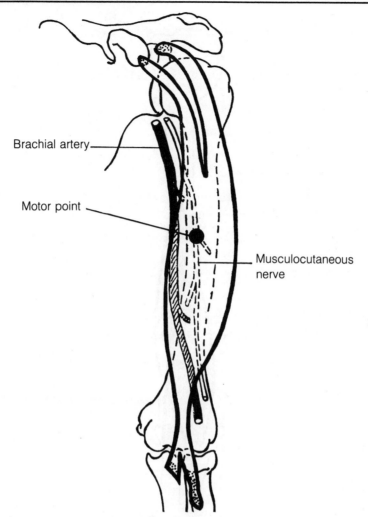

Brachial artery

Motor point

Musculocutaneous nerve

ORIGIN: Short head from tip of coracoid process of scapula, long head from supraglenoid tuberosity of scapula

INSERTION: Radial tuberosity and by lacertus fibrosus to origins of forearm flexors

FUNCTION: Flexes and supinates forearm, flexes arm when forearm is fixed

NERVE: Musculocutaneous

ARTERY: Muscular branches of brachial

References

	GRAY	GRANT'S ATLAS	NETTER
Muscle	527	6-17, 6-43, 6-46, 6-57	401, 406, 408
Nerve	528, 1205, 1207, 1212	6-26	405, 406, 446, 447
Artery	719	6-26, 6-43	408

BRACHIALIS

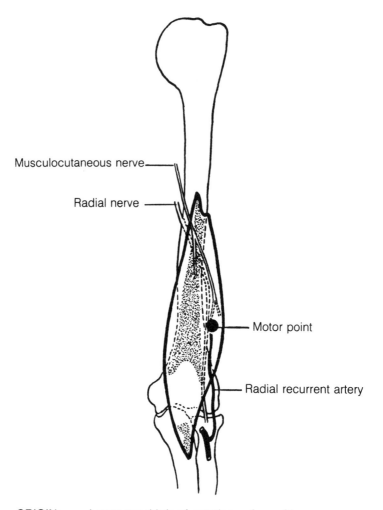

Musculocutaneous nerve

Radial nerve

Motor point

Radial recurrent artery

ORIGIN: Lower two-thirds of anterior surface of humerus
INSERTION: Coronoid process and tuberosity of ulna
FUNCTION: Flexes forearm
NERVE: Musculocutaneous, radial (may be afferent)
ARTERY: Radial recurrent, brachial

References

	GRAY	GRANT'S ATLAS	NETTER
Muscle	528	6-39, 6-59	406, 422
Nerve	528, 1205, 1207, 1212, 1220	6-26	405, 406, 422, 447
Artery	719, 726	6-43	408

TRICEPS BRACHII

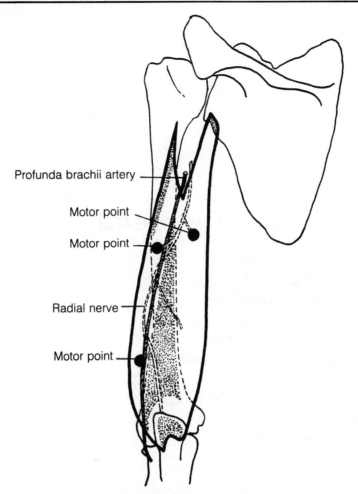

Profunda brachii artery —

Motor point —

Motor point —

Radial nerve —

Motor point —

ORIGIN: Long head from infraglenoid tuberosity of scapula, lateral head from posterior and lateral surface of humerus, medial head from lower posterior surface of humerus

INSERTION: Upper posterior surface of olecranon and deep fascia of forearm

FUNCTION: Extends forearm; if arm is abducted, long head aids in adducting it

NERVE: Radial

ARTERY: Branch of profunda brachii

References

	GRAY	GRANT'S ATLAS	NETTER
Muscle	528	6-40, 6-43, 6-64	399, 402, 407
Nerve	529, 1205, 1207, 1220	6-41	405, 407, 450
Artery	718	6-26, 6-41	407

5. VOLAR MUSCLES OF THE FOREARM

Superficial Group
    Pronator teres
    Flexor carpi radialis
    Palmaris longus
    Flexor carpi ulnaris
    Flexor digitorum superficialis
Deep Group
    Flexor digitorum profundus
    Flexor pollicis longus
    Pronator quadratus

PRONATOR TERES

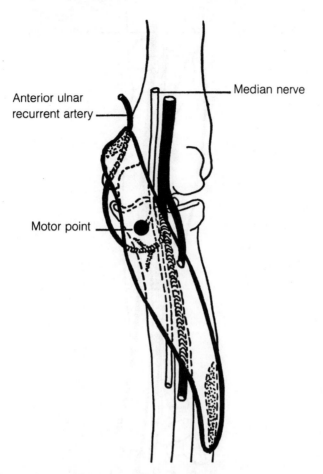

Anterior ulnar recurrent artery

Median nerve

Motor point

ORIGIN: <u>Humeral head</u> from medial epicondylar ridge of humerus and common flexor tendon, <u>ulnar head</u> from medial side of coronoid process of ulna

INSERTION: Middle of lateral surface of radius

FUNCTION: Pronates forearm, assists in flexing forearm

NERVE: Median

ARTERY: Anterior ulnar recurrent

References

	GRAY	GRANT'S ATLAS	NETTER
Muscle	530	6-79	406, 414
Nerve	531, 1205, 1207, 1213	6-81	404, 405, 408, 448
Artery	720, 726	Not shown	421

FLEXOR CARPI RADIALIS

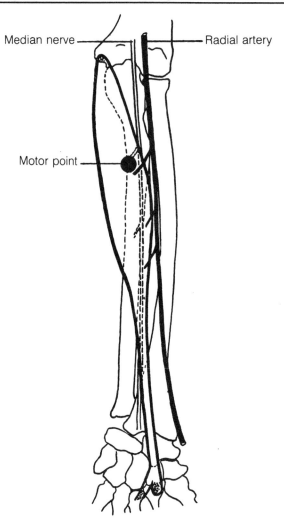

Median nerve — Radial artery

Motor point

ORIGIN: Common flexor tendon from medial epicondyle of humerus, fascia of forearm

INSERTION: Base of 2d and 3d metacarpal bones

FUNCTION: Flexes wrist, assists in pronating and abducting hand, assists in flexing forearm

NERVE: Median

ARTERY: Muscular branches of radial

References

	GRAY	GRANT'S ATLAS	NETTER
Muscle	531	6-79	416, 420
Nerve	531, 1205, 1207, 1213	Not shown	405, 448
Artery	726	Not shown	421

PALMARIS LONGUS

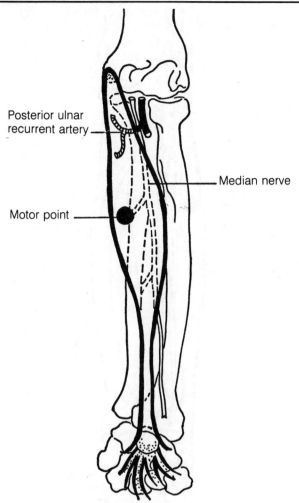

Posterior ulnar recurrent artery

Median nerve

Motor point

ORIGIN: Common flexor tendon from medial epicondyle of humerus
INSERTION: Transverse carpal ligament and palmar aponeurosis
FUNCTION: Flexes wrist, assists in pronation and flexion of forearm
NERVE: Median
ARTERY: Posterior ulnar recurrent

References

	GRAY	GRANT'S ATLAS	NETTER
Muscle	531	6-79, 6-87	416, 420
Nerve	532, 1205, 1207, 1213	Not shown	405, 448
Artery	720	Not shown	422

FLEXOR CARPI ULNARIS

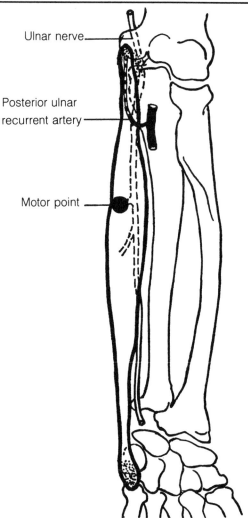

Ulnar nerve

Posterior ulnar
recurrent artery

Motor point

ORIGIN: Humeral head from common flexor tendon from medial epicondyle of humerus, ulnar head from olecranon and dorsal border
of ulna

INSERTION: Pisiform, hamate, 5th metacarpal bones

FUNCTION: Flexes wrist and assists in adducting it; assists in flexing forearm

NERVE: Ulnar

ARTERY: Posterior ulnar recurrent

References

	GRAY	GRANT'S ATLAS	NETTER
Muscle	532	6-79, 6-81	416, 420
Nerve	532, 1205, 1207, 1217	6-81, 6-83	405, 449
Artery	720	6-66, 6-83	422

FLEXOR DIGITORUM SUPERFICIALIS

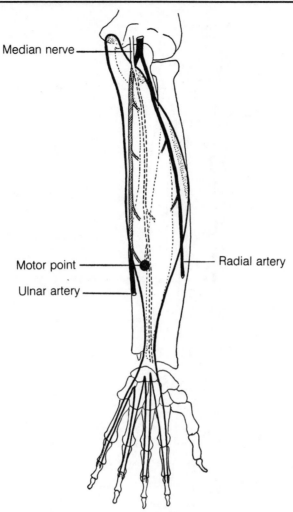

Median nerve

Motor point

Radial artery

Ulnar artery

ORIGIN: Humeral head from common flexor tendon from medial epicondyle of humerus, ulnar head from coronoid process of ulna, radial head from oblique line of radius

INSERTION: Margins of palmar surface of middle phalanx of medial 4 digits

FUNCTION: Flexes middle and proximal phalanges of medial 4 digits, aids in flexing wrist and forearm

NERVE: Median

ARTERY: Muscular branches of ulnar, muscular branches of radial

References

	GRAY	GRANT'S ATLAS	NETTER
Muscle	532	6-79, 6-87, 6-99	417, 421, 431
Nerve	533, 1205, 1207, 1213	Not shown	405, 448
Artery	721, 726	6-81	421

FLEXOR DIGITORUM PROFUNDUS

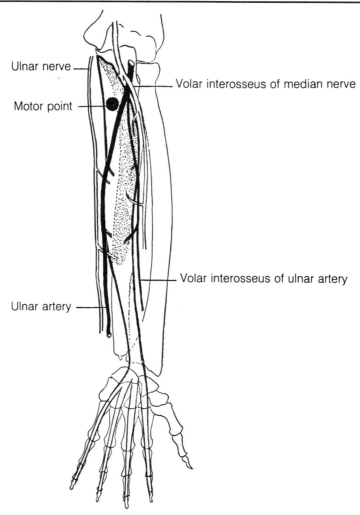

Ulnar nerve

Volar interosseus of median nerve

Motor point

Volar interosseus of ulnar artery

Ulnar artery

ORIGIN: Medial and anterior surface of ulna, interosseus membrane, deep fascia of forearm

INSERTION: Distal phalanges of medial 4 digits

FUNCTION: Flexes terminal phalanges of medial 4 digits after superficialis flexes 2d phalanges, aids in flexing wrist

NERVE: Ulnar, volar interosseus of median

ARTERY: Volar interosseus of ulnar, muscular branches of ulnar

References

	GRAY	GRANT'S ATLAS	NETTER
Muscle	533	6-81, 6-93, 6-99	417, 422, 431
Nerve	534, 1205, 1207, 1213, 1217	6-81, 6-83	405, 448, 449
Artery	721	6-83	422

FLEXOR POLLICIS LONGUS

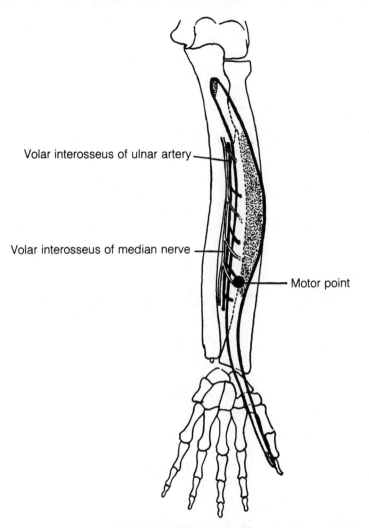

Volar interosseus of ulnar artery

Volar interosseus of median nerve

Motor point

ORIGIN: Volar surface of radius, adjacent interosseus membrane, medial border of coronoid process of ulna

INSERTION: Base of distal phalanx of thumb on palmar surface

FUNCTION: Flexes thumb

NERVE: Volar interosseus of median

ARTERY: Volar interosseus of ulnar

References

	GRAY	GRANT'S ATLAS	NETTER
Muscle	535	6-83, 6-95	417, 421
Nerve	535, 1205, 1207, 1213	6-83	448
Artery	721	6-83	422

PRONATOR QUADRATUS

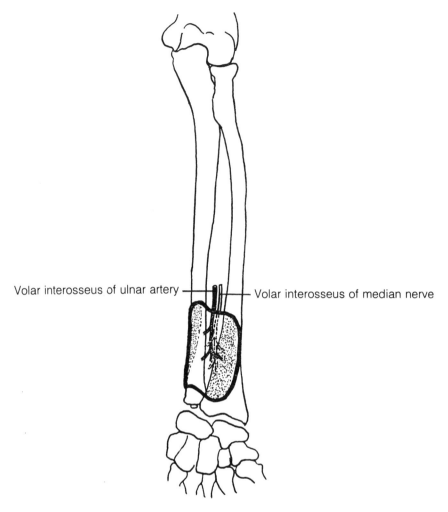

Volar interosseus of ulnar artery — — Volar interosseus of median nerve

ORIGIN: Distal fourth of volar surface of ulna
INSERTION: Distal fourth of lateral border on volar surface of radius
FUNCTION: Pronates forearm
NERVE: Volar interosseus of median
ARTERY: Volar interosseus of ulnar

References

	GRAY	GRANT'S ATLAS	NETTER
Muscle	535	6-99	414, 422, 434
Nerve	535, 1205, 1207, 1213	Not shown	Not shown
Artery	721	Not shown	422

6. DORSAL MUSCLES OF THE FOREARM

Superficial Group
Brachioradialis
Extensor carpi radialis longus
Extensor carpi radialis brevis
Extensor digitorum
Extensor digiti minimi
Extensor carpi ulnaris
Anconeus

Deep Group
Supinator
Abductor pollicis longus
Extensor pollicis brevis
Extensor pollicis longus
Extensor indicis

BRACHIORADIALIS

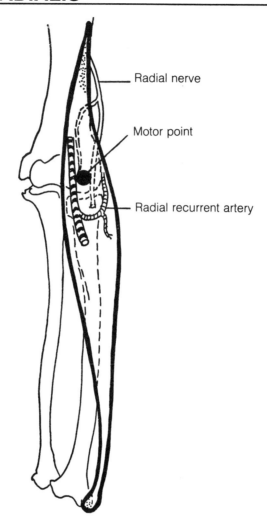

Radial nerve

Motor point

Radial recurrent artery

ORIGIN: Proximal two-thirds of lateral supracondylar ridge of humerus, lateral intermuscular septum

INSERTION: Lateral side of base of styloid process of radius

FUNCTION: Flexes forearm after flexion has been started by biceps and brachialis; may also act as a semipronator and semisupinator

NERVE: Radial

ARTERY: Radial recurrent

References

	GRAY	GRANT'S ATLAS	NETTER
Muscle	535	6-79	418, 420
Nerve	536, 1205, 1207, 1220	6-83	421
Artery	726	6-81, 6-83	408, 421

EXTENSOR CARPI RADIALIS LONGUS

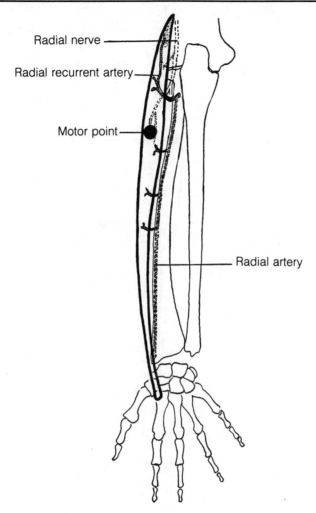

Radial nerve

Radial recurrent artery

Motor point

Radial artery

ORIGIN: Lower third of lateral supracondylar ridge of humerus, lateral intermuscular septum

INSERTION: Dorsal surface of base of 2d metacarpal bone

FUNCTION: Extends wrist, abducts hand

NERVE: Radial

ARTERY: Muscular branches of radial, radial recurrent

References

	GRAY	GRANT'S ATLAS	NETTER
Muscle	536	6-102, 6-103, 6-105, 6-107	415, 419, 440
Nerve	536, 1205, 1207, 1220	Not shown	451
Artery	726	6-83	Not shown

EXTENSOR CARPI RADIALIS BREVIS

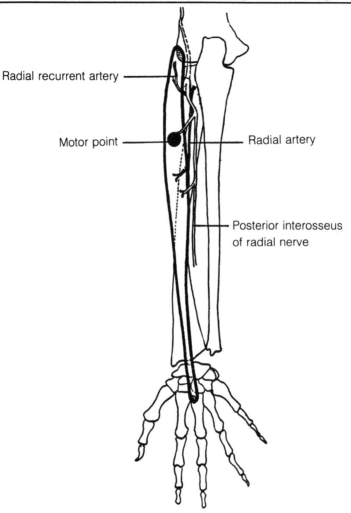

Radial recurrent artery

Motor point

Radial artery

Posterior interosseus of radial nerve

ORIGIN: From common extensor tendon from lateral epicondyle of humerus, radial collateral ligament of elbow joint, intermuscular septa

INSERTION: Dorsal surface of base of 3d metacarpal bone

FUNCTION: Extends wrist, abducts hand

NERVE: Posterior interosseus of radial

ARTERY: Muscular branches of radial, radial recurrent

References

	GRAY	GRANT'S ATLAS	NETTER
Muscle	536	6-103, 6-107	415, 419, 420
Nerve	537, 1205, 1207, 1221	6-83	451
Artery	726	6-83	Not shown

EXTENSOR DIGITORUM

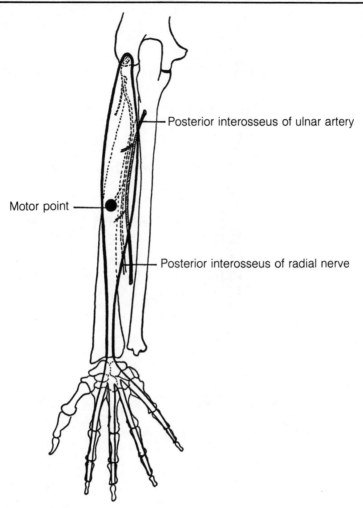

Posterior interosseus of ulnar artery

Motor point

Posterior interosseus of radial nerve

ORIGIN: Lateral epicondyle of humerus by common extensor tendon, intermuscular septa

INSERTION: Lateral and dorsal surface of phalanges of medial 4 digits

FUNCTION: Extends medial 4 digits; assists in extension of wrist

NERVE: Posterior interosseus of radial

ARTERY: Posterior interosseus of ulnar

References

	GRAY	GRANT'S ATLAS	NETTER
Muscle	537	6-103, 6-107	415, 418, 443
Nerve	537, 1205, 1207, 1221	6-114	451
Artery	721	6-114	419

EXTENSOR DIGITI MINIMI

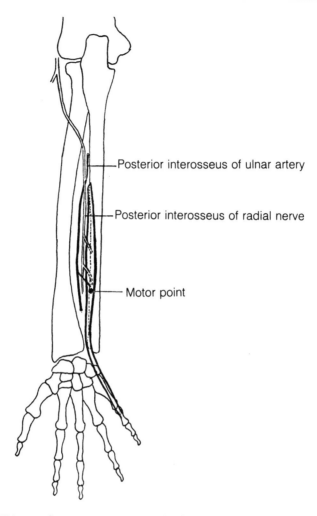

Posterior interosseus of ulnar artery

Posterior interosseus of radial nerve

Motor point

ORIGIN: Common extensor tendon from lateral epicondyle of the humerus, intermuscular septa

INSERTION: Dorsum of proximal phalanx of 5th digit

FUNCTION: Extends 5th digit

NERVE: Posterior interosseus of radial

ARTERY: Posterior interosseus of ulnar

References

	GRAY	GRANT'S ATLAS	NETTER
Muscle	537	6-103, 6-107	415, 418, 443
Nerve	538, 1205, 1207, 1221	6-114	451
Artery	721	6-114	419

EXTENSOR CARPI ULNARIS

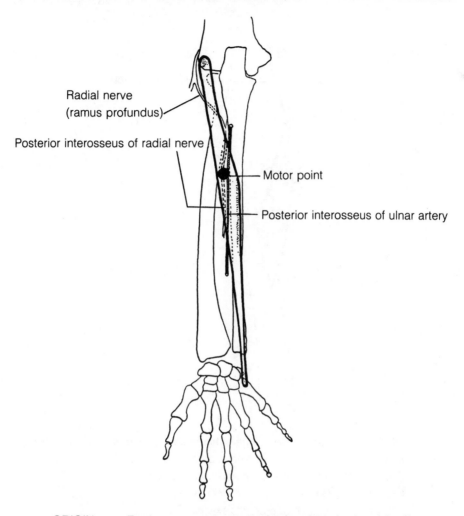

Radial nerve
(ramus profundus)

Posterior interosseus of radial nerve

Motor point

Posterior interosseus of ulnar artery

ORIGIN: From common extensor tendon from lateral epicondyle of humerus, and from posterior border of ulna

INSERTION: Medial side of base of 5th metacarpal bone

FUNCTION: Extends wrist, adducts hand

NERVE: Posterior interosseus of radial

ARTERY: Posterior interosseus of ulnar

References

	GRAY	GRANT'S ATLAS	NETTER
Muscle	538	6-103	415, 443
Nerve	538, 1205, 1207, 1221	6-114	451
Artery	721	6-114	419

ANCONEUS

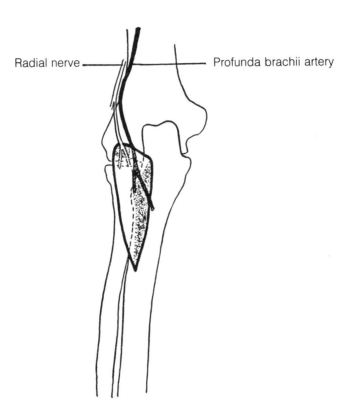

Radial nerve — Profunda brachii artery

ORIGIN: Lateral epicondyle of humerus, posterior ligament of elbow joint
INSERTION: Lateral side of olecranon and posterior surface of ulna
FUNCTION: Assists triceps in extending forearm
NERVE: Radial
ARTERY: Branch of profunda brachii

References

	GRAY	GRANT'S ATLAS	NETTER
Muscle	538	6-66, 6-102, 6-114	407, 418
Nerve	538, 1205, 1207, 1220	6-103	Not shown
Artery	717	Not shown	Not shown

SUPINATOR

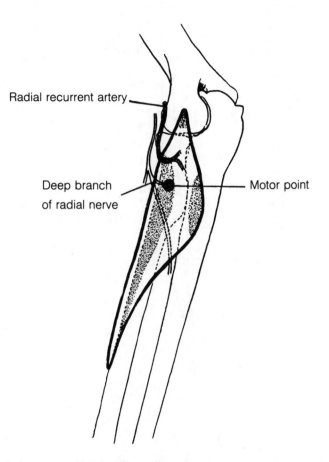

Radial recurrent artery

Deep branch
of radial nerve

Motor point

ORIGIN: Lateral epicondyle of humerus, lateral ligament of elbow joint and annular ligament of radius, supinator crest and fossa of ulna

INSERTION: Lateral and anterior surface of radius in its upper third

FUNCTION: Supinates forearm

NERVE: Deep branch of radial

ARTERY: Radial recurrent, posterior interosseous of ulnar

References

	GRAY	GRANT'S ATLAS	NETTER
Muscle	538	6-59, 6-114	414, 421
Nerve	539, 1205, 1207, 1221	6-83	451
Artery	721, 726	6-83	419

ABDUCTOR POLLICIS LONGUS

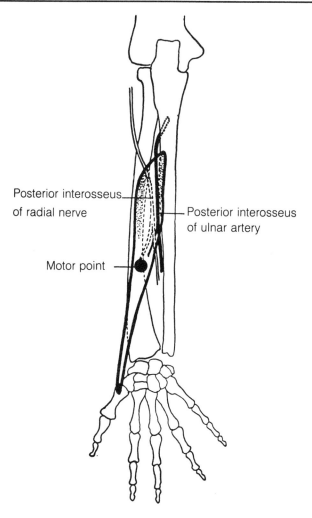

Posterior interosseus of radial nerve

Posterior interosseus of ulnar artery

Motor point

ORIGIN: Posterior surface of ulna, interosseus membrane, middle third of posterior surface of radius

INSERTION: Radial side of base of 1st metacarpal bone

FUNCTION: Abducts thumb and wrist

NERVE: Posterior interosseus of radial

ARTERY: Posterior interosseus of ulnar

References

	GRAY	GRANT'S ATLAS	NETTER
Muscle	539	6-90, 6-103, 6-114	415, 419, 440
Nerve	539, 1205, 1207, 1221	6-114	451
Artery	721	6-114	419

EXTENSOR POLLICIS BREVIS

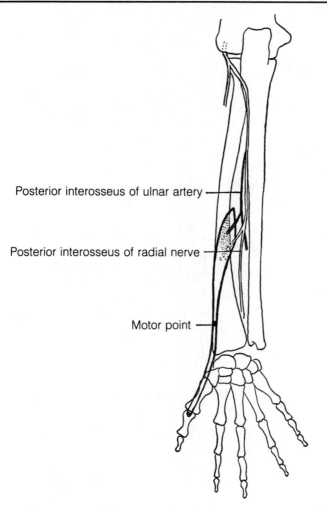

Posterior interosseus of ulnar artery

Posterior interosseus of radial nerve

Motor point

ORIGIN: Posterior surface of radius, interosseus membrane
INSERTION: Base of proximal phalanx of thumb
FUNCTION: Extends proximal phalanx of thumb
NERVE: Posterior interosseus of radial
ARTERY: Posterior interosseus of ulnar

References

	GRAY	GRANT'S ATLAS	NETTER
Muscle	540	6-103, 6-114	415, 418, 419
Nerve	540, 1205, 1207, 1221	6-114	451
Artery	721	6-114	419

EXTENSOR POLLICIS LONGUS

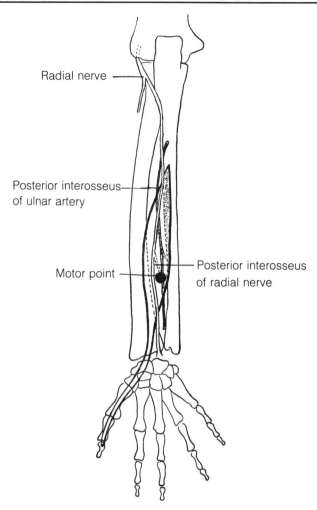

Radial nerve

Posterior interosseus of ulnar artery

Motor point

Posterior interosseus of radial nerve

ORIGIN: Middle third of posterior surface of ulna, interosseus membrane
INSERTION: Base of distal phalanx of thumb
FUNCTION: Extends terminal phalanx of thumb
NERVE: Posterior interosseus of radial
ARTERY: Posterior interosseus of ulnar

References

	GRAY	GRANT'S ATLAS	NETTER
Muscle	540	6-103, 6-114	415, 418, 440
Nerve	540, 1205, 1207, 1221	6-114	451
Artery	721	6-114	419

EXTENSOR INDICIS

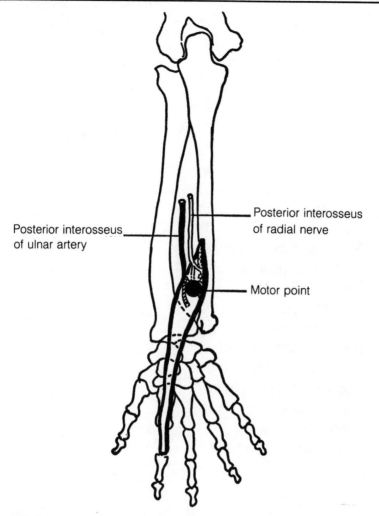

Posterior interosseus
of ulnar artery

Posterior interosseus
of radial nerve

Motor point

ORIGIN:	Posterior surface of ulna, interosseus membrane
INSERTION:	Dorsum of proximal phalanx of index finger
FUNCTION:	Extends proximal phalanx of index finger
NERVE:	Posterior interosseus of radial
ARTERY:	Posterior interosseus of ulnar

References

	GRAY	GRANT'S ATLAS	NETTER
Muscle	540	6-107, 6-114	415, 419, 443
Nerve	540, 1205, 1207, 1221	6-114	451
Artery	721	6-114	419

7. MUSCLES OF THE HAND

Abductor pollicis brevis
Opponens pollicis
Flexor pollicis brevis
Adductor pollicis
Palmaris brevis
Abductor digiti minimi
Flexor digiti minimi brevis
Opponens digiti minimi
Lumbricales
Interossei dorsales
Interossei palmares

ABDUCTOR POLLICIS BREVIS

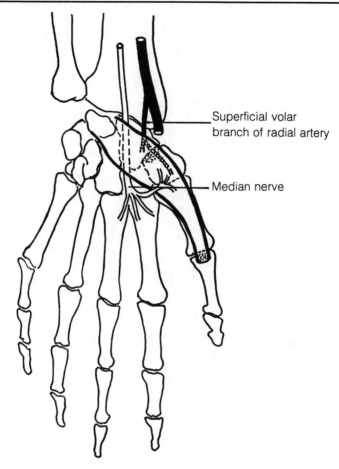

Superficial volar branch of radial artery

Median nerve

ORIGIN: Transverse carpal ligament, scaphoid and trapezium bones
INSERTION: Radial side of base of proximal phalanx of thumb
FUNCTION: Abducts thumb, draws thumb forward at right angles to palm
NERVE: Muscular branches of median
ARTERY: Superficial volar branch of radial

References

	GRAY	GRANT'S ATLAS	NETTER
Muscle	550	6-90	439
Nerve	550, 1205, 1207, 1216	6-90	439, 448
Artery	727	6-90	439

OPPONENS POLLICIS

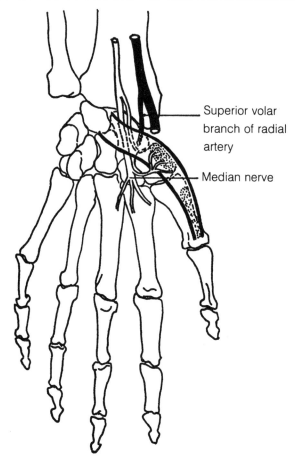

Superior volar branch of radial artery

Median nerve

ORIGIN: Transverse carpal ligament, trapezium bone
INSERTION: Anterior surface, radial side of 1st metacarpal bone
FUNCTION: Draws 1st metacarpal bone forward, and medially, opposing thumb
            to each of the other digits
NERVE: Muscular branches of median
ARTERY: Superficial volar branch of radial

References

	GRAY	GRANT'S ATLAS	NETTER
Muscle	550	6-83, 6-91	438
Nerve	550, 1205, 1207, 1216	6-91	438, 448
Artery	727	6-84	439

FLEXOR POLLICIS BREVIS

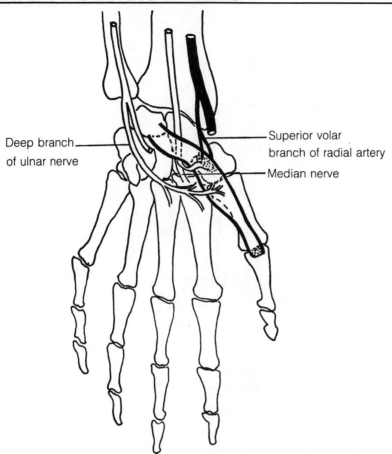

Deep branch
of ulnar nerve

Superior volar
branch of radial artery

Median nerve

ORIGIN: Transverse carpal ligament, trapezium bone
INSERTION: Base of proximal phalanx of thumb
FUNCTION: Flexes proximal phalanx of thumb
NERVE: Muscular branches of median; deep branch of ulnar
ARTERY: Superficial volar branch of radial

References

	GRAY	GRANT'S ATLAS	NETTER
Muscle	550	6-83, 6-91	434, 438, 449
Nerve	551, 1205, 1207, 1216, 1219	6-91	438, 448
Artery	727	6-90	439

ADDUCTOR POLLICIS

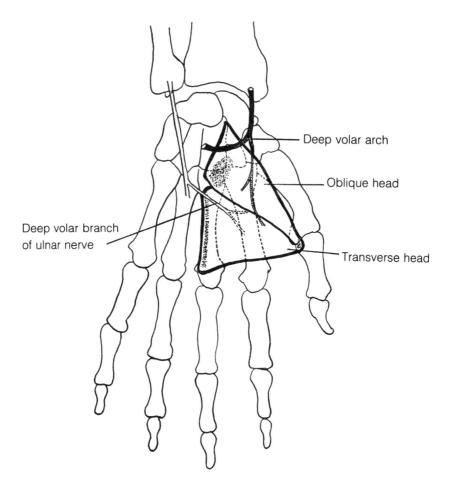

Deep volar arch

Oblique head

Deep volar branch
of ulnar nerve

Transverse head

ORIGIN: Oblique head from trapezium, trapezoid and capitate bones, base
            of 2d and 3d metacarpal bone, transverse head from volar sur-
            face of 3d metacarpal bone
INSERTION: Ulnar side of base of proximal phalanx of thumb
FUNCTION: Adducts thumb, aids in opposition
NERVE: Deep volar branch of ulnar
ARTERY: Deep volar arch

References

	GRAY	GRANT'S ATLAS	NETTER
Muscle	551	6-88, 6-95, 6-116	433, 434, 438
Nerve	552, 1205, 1207, 1219	6-95	438, 449
Artery	728	6-95	438, 439

PALMARIS BREVIS

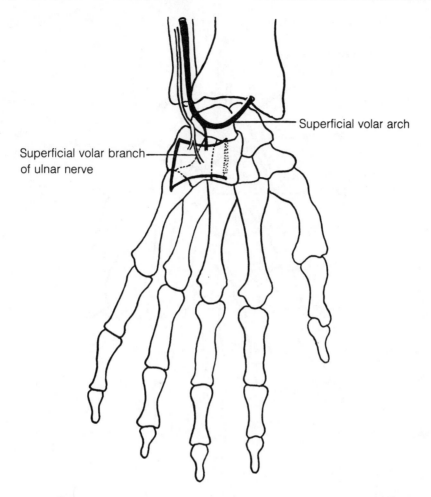

Superficial volar arch

Superficial volar branch of ulnar nerve

ORIGIN: Ulnar side of transverse carpal ligament, palmar aponeurosis
INSERTION: Skin on ulnar border of palm
FUNCTION: Corrugates skin on ulnar side of palm, deepens the hollow of the hand
NERVE: Superficial volar branch of ulnar
ARTERY: Superficial volar arch

References

	GRAY	GRANT'S ATLAS	NETTER
Muscle	552	6-90	432, 449
Nerve	552, 1205, 1207, 1218	Not shown	432, 449
Artery	727	Not shown	439

ABDUCTOR DIGITI MINIMI

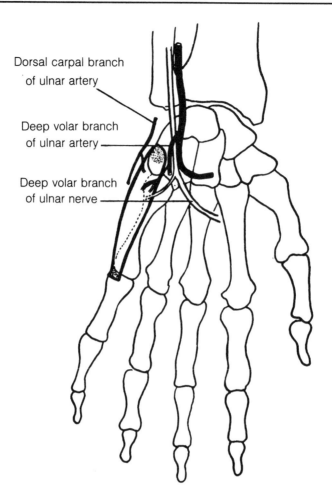

Dorsal carpal branch of ulnar artery

Deep volar branch of ulnar artery

Deep volar branch of ulnar nerve

ORIGIN: Pisiform bone, tendon of flexor carpi ulnaris

INSERTION: Medial side of base of proximal phalanx of 5th digit and aponeurosis of extensor digiti minimi

FUNCTION: Abducts 5th digit from 4th digit

NERVE: Deep volar branch of ulnar

ARTERY: Deep volar branch of ulnar, dorsal carpal branch of ulnar

References

	GRAY	GRANT'S ATLAS	NETTER
Muscle	552	6-90, 6-119	434, 438
Nerve	553, 1205, 1207, 1218	6-91	434, 449
Artery	722	6-95, 6-119	439

FLEXOR DIGITI MINIMI BREVIS

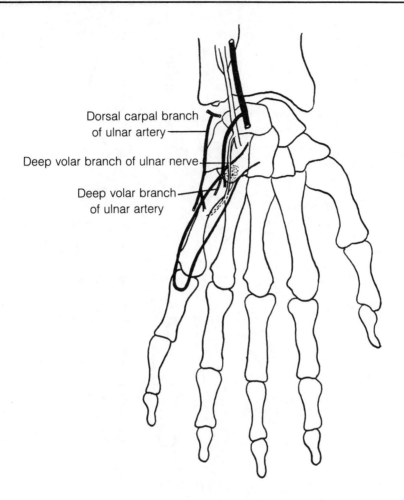

Dorsal carpal branch of ulnar artery

Deep volar branch of ulnar nerve

Deep volar branch of ulnar artery

ORIGIN: Transverse carpal ligament, hamulus of hamate bone
INSERTION: Ulnar side, base of proximal phalanx of 5th digit
FUNCTION: Flexes proximal phalanx of 5th digit
NERVE: Deep volar branch of ulnar
ARTERY: Deep volar branch of ulnar, dorsal carpal branch of ulnar

References

	GRAY	GRANT'S ATLAS	NETTER
Muscle	553	6-91	434, 438
Nerve	554, 1205, 1207, 1218	6-91	449
Artery	722	6-95	Not shown

OPPONENS DIGITI MINIMI

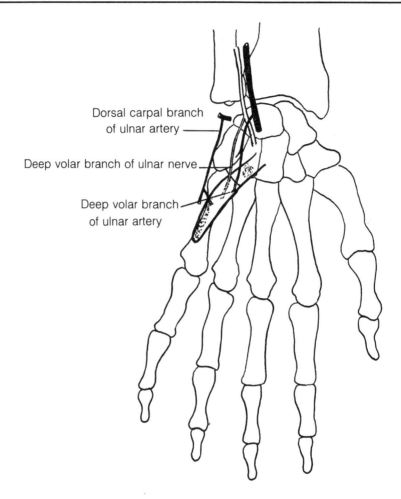

Dorsal carpal branch
of ulnar artery

Deep volar branch of ulnar nerve

Deep volar branch
of ulnar artery

ORIGIN: Transverse carpal ligament, hamulus of hamate bone
INSERTION: Ulnar margin of 5th metacarpal bone
FUNCTION: Draws 5th metacarpal bone forward to face thumb, deepens hollow of hand
NERVE: Deep volar branch of ulnar
ARTERY: Deep volar branch of ulnar, dorsal carpal branch of ulnar

References

	GRAY	GRANT'S ATLAS	NETTER
Muscle	554	6-83, 6-91, 6-119	434, 438
Nerve	554, 1205, 1207, 1218	6-91	449
Artery	722	6-95, 6-119	Not shown

LUMBRICALES

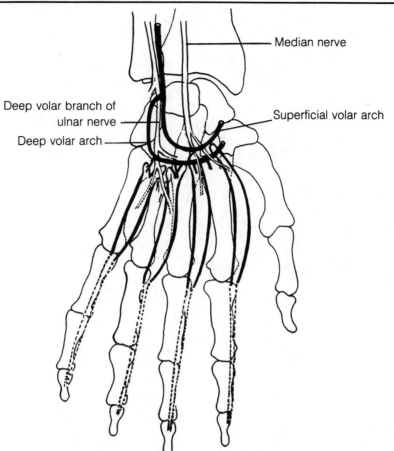

Median nerve

Deep volar branch of
ulnar nerve

Deep volar arch

Superficial volar arch

ORIGIN: There are 4 lumbricales, all arising from tendons of flexor digitorum profundus: 1st from radial side of tendon for index finger, 2d from radial side of tendon for middle finger, 3d from adjacent sides of tendons for middle and ring fingers, 4th from adjacent sides of tendons for ring and little fingers

INSERTION: With tendons of extensor digitorum and interossei into bases of terminal phalanges of medial 4 digits

FUNCTION: Flex fingers at metacarpophalangeal joints, extend fingers at interphalangeal joints

NERVE: Median, to lateral two muscles, deep volar branch of ulnar to medial two muscles

ARTERY: Superficial and deep volar arches

References

	GRAY	GRANT'S ATLAS	NETTER
Muscle	554	6-83, 6-91	434, 436
Nerve	554, 1205, 1207, 1216, 1218	6-91	448, 449
Artery	723, 728	6-90	439

INTEROSSEI DORSALES

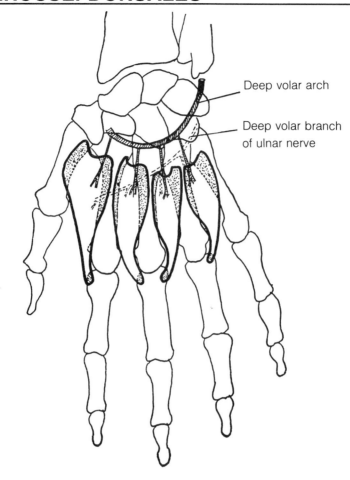

Deep volar arch

Deep volar branch
of ulnar nerve

ORIGIN: There are 4 dorsal interossei; each arises by 2 heads from adjacent sides of metacarpal bones

INSERTION: 1st into radial side of proximal phalanx of 2d digit, 2d into radial
side of proximal phalanx of 3d digit, 3d into ulnar side of proximal
phalanx of 3d digit, 4th into ulnar side of proximal phalanx of 4th
digit

FUNCTION: Abduct index, middle and ring fingers from the mid line of the
hand

NERVE: Deep volar branch of ulnar

ARTERY: Deep volar arch

References

	GRAY	GRANT'S ATLAS	NETTER
Muscle	554	6-99, 6-103, 6-107	433, 437, 438
Nerve	555, 1205, 1207, 1218	6-99	449
Artery	728	6-98, 6-116C	438, 439

INTEROSSEI PALMARES

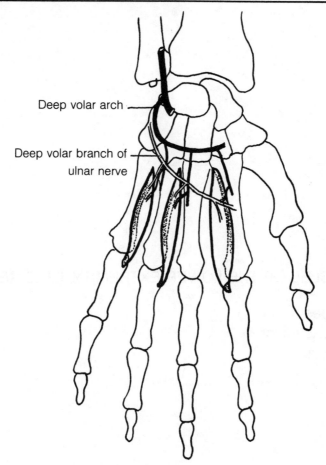

Deep volar arch

Deep volar branch of ulnar nerve

ORIGIN: There are 3 volar interossei: 1st from ulnar side of 2d metacarpal bone, 2d from radial side of 4th metacarpal bone, 3d from radial side of 5th metacarpal bone

INSERTION: 1st into ulnar side of proximal phalanx of 2d digit, 2d into radial side of proximal phalanx of 4th digit, 3d into radial side of proximal phalanx of 5th digit

FUNCTION: Each muscle adducts digit into which it is inserted toward middle digit

NERVE: Deep volar branch of ulnar

ARTERY: Deep volar arch

References

	GRAY	GRANT'S ATLAS	NETTER
Muscle	555	6-99	438
Nerve	555, 1205, 1207, 1218	6-99	449
Artery	728	6-98	439

8.   MUSCLES OF THE LOWER EXTREMITY, ILIAC REGION

Psoas major
Psoas minor
Iliacus

PSOAS MAJOR

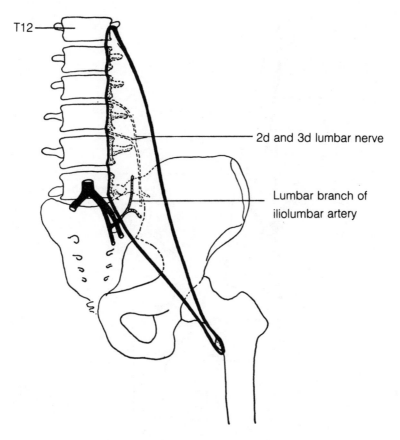

T12

2d and 3d lumbar nerve

Lumbar branch of
iliolumbar artery

ORIGIN: Anterior surface of transverse processes and bodies of lumbar
vertebrae, corresponding intervertebral discs
INSERTION: Lesser trochanter of femur (with iliacus forms iliopsoas)
FUNCTION: Flexes thigh, flexes vertebral column on pelvis when leg is fixed
NERVE: 2d and 3d lumbar
ARTERY: Lumbar branch of iliolumbar

References

	GRAY	GRANT'S ATLAS	NETTER
Muscle	557	5-14, 5-17, 5-21	181, 246, 466
Nerve	557, 1226	Not shown	467
Artery	756	Not shown	247

PSOAS MINOR

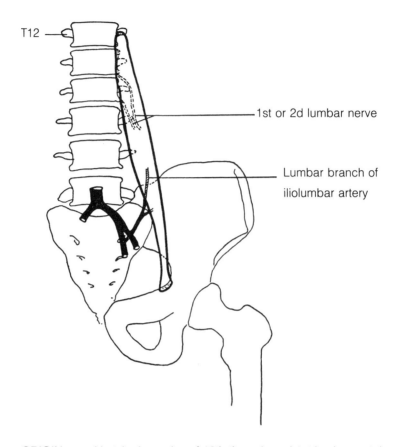

T12

1st or 2d lumbar nerve

Lumbar branch of
iliolumbar artery

ORIGIN: Vertebral margins of 12th thoracic and 1st lumbar vertebra, cor-
responding intervertebral disc
INSERTION: Pectineal line, iliopectineal eminence
FUNCTION: Flexes pelvis on vertebral column, assists psoas major in flexing
vertebral column on pelvis. This muscle is inconstant, absent in
40% of bodies
NERVE: 1st or 2d lumbar
ARTERY: Lumbar branch of iliolumbar

References

	GRAY	GRANT'S ATLAS	NETTER
Muscle	557	5-20	246, 466
Nerve	558, 1226	Not shown	467
Artery	756	Not shown	247

ILIACUS*

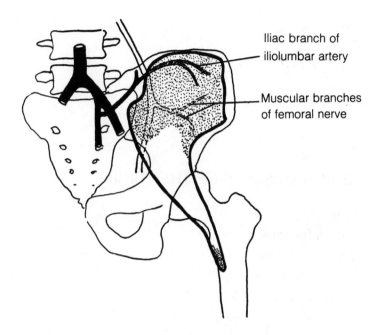

Iliac branch of
iliolumbar artery

Muscular branches
of femoral nerve

ORIGIN: Upper two-thirds of iliac fossa; iliac crest; anterior sacroiliac, lumbosacral, and iliolumbar ligaments; ala of sacrum

INSERTION: Tendon of psoas major, lesser trochanter, capsule of hip joint, body of femur (with psoas major forms iliopsoas)

FUNCTION: Flexes thigh, tilts pelvis forward when leg is fixed

NERVE: Muscular branches of femoral

ARTERY: Iliac branch of iliolumbar, superior gluteal

References

	GRAY	GRANT'S ATLAS	NETTER
Muscle	558	5-21	247, 466, 471
Nerve	558, 1226, 1232	2-103	466, 467
Artery	756, 757	Not shown	247, 324

*With psoas major (and minor) forms iliopsoas

9. ANTERIOR MUSCLES OF THE THIGH

Sartorius
Rectus femoris
Vastus lateralis
Vastus medialis
Vastus intermedius

SARTORIUS

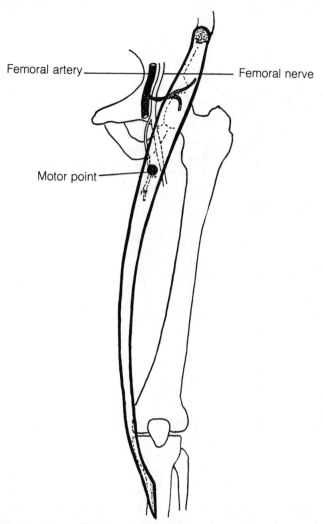

Femoral artery ———————— Femoral nerve

Motor point ————

ORIGIN: Anterior superior iliac spine, upper half of iliac notch
INSERTION: Upper part of medial surface of tibia
FUNCTION: Flexes leg on thigh, flexes thigh on pelvis, rotates thigh laterally
NERVE: Muscular branches of femoral
ARTERY: Muscular branches of femoral

References

	GRAY	GRANT'S ATLAS	NETTER
Muscle	561	5-16, 5-20	462, 470, 488
Nerve	562, 1226, 1232	5-24	467, 470
Artery	765	Not shown	245, 470

RECTUS FEMORIS

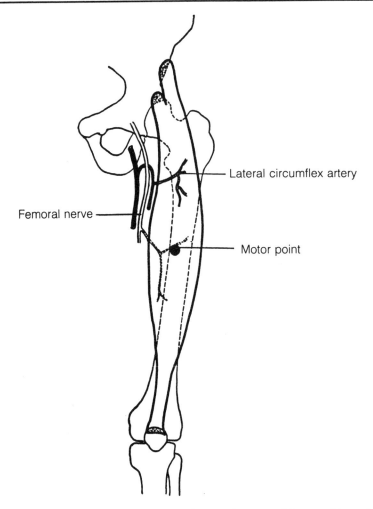

Lateral circumflex artery

Femoral nerve

Motor point

ORIGIN: (Rectus femoris is 1 of 4 muscles comprising quadriceps femoris.) Straight head from anterior inferior iliac spine, reflected head from groove on upper brim of acetabulum

INSERTION: Upper border of patella; by ligamentum patellae into tibial tuberosity

FUNCTION: Extends leg, flexes thigh

NERVE: Muscular branches of femoral

ARTERY: Lateral femoral circumflex

References

	GRAY	GRANT'S ATLAS	NETTER
Muscle	562-563	5-20	462, 470
Nerve	563, 1226, 1233	5-24	467, 470
Artery	767	5-24	471

VASTUS LATERALIS

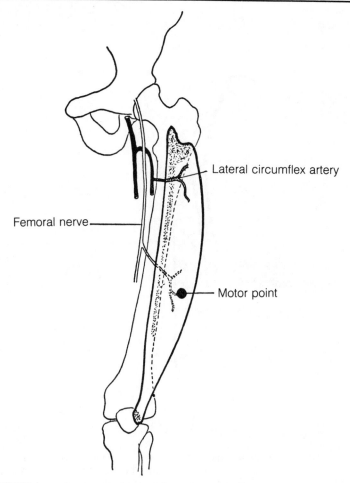

Lateral circumflex artery

Femoral nerve

Motor point

ORIGIN: (Vastus lateralis is 1 of 4 muscles comprising quadriceps femoris.) Capsule of hip joint, intertrochanteric line, greater trochanter, gluteal tuberosity, linea aspera, lateral intermuscular septum

INSERTION: Lateral border by patella, by ligamentum patellae into tibial tuberosity

FUNCTION: Extends leg

NERVE: Muscular branches of femoral

ARTERY: Lateral femoral circumflex, lateral superior genicular

References

	GRAY	GRANT'S ATLAS	NETTER
Muscle	562-563	5-20	462, 464
Nerve	563, 1226, 1233	5-24	467, 471
Artery	767, 772	5-24	471, 490

VASTUS MEDIALIS

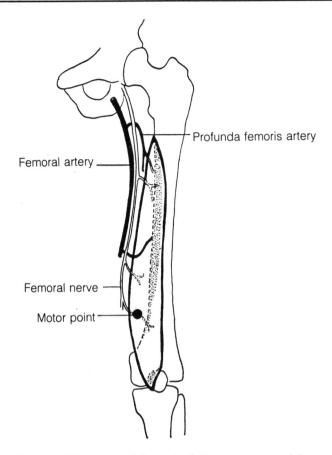

Profunda femoris artery

Femoral artery

Femoral nerve

Motor point

ORIGIN: (Vastus medialis is 1 of 4 muscles comprising quadriceps femoris.) Lower half of intertrochanteric line, linea aspera medial supracondylar line, medial intermuscular septum, tendon of adductor magnus

INSERTION: Quadriceps femoris tendon, medial border of patella, capsule of knee joint, by ligamentum patellae into tibial tuberosity

FUNCTION: Extends leg and draws patella medially

NERVE: Muscular branches of femoral

ARTERY: Muscular branches of femoral, muscular branches of profunda femoris, genicular branches of popliteal

References

	GRAY	GRANT'S ATLAS	NETTER
Muscle	562-563	5-20	462, 470
Nerve	563, 1226, 1233	5-24	467, 471
Artery	765, 769, 770	Not shown	471, 481, 485

VASTUS INTERMEDIUS*

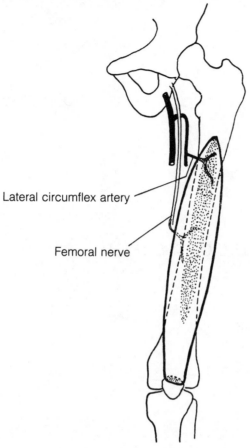

Lateral circumflex artery

Femoral nerve

ORIGIN: (Vastus intermedius is 1 of 4 muscles comprising quadriceps femoris.) Upper two-thirds of anterior and lateral surface of femur, lower half of linea aspera, upper part of lateral supracondylar line, lateral intermuscular septum

INSERTION: Deep surface of tendons of rectus and vasti muscles, by ligamentum patellae into tibial tuberosity

FUNCTION: Extends leg

NERVE: Muscular branches of femoral

ARTERY: Lateral femoral circumflex

References

	GRAY	GRANT'S ATLAS	NETTER
Muscle	562-563	5-21, 5-67	462, 471, 477, 480
Nerve	563, 1226, 1233	5-24	467, 471
Artery	767	5-24	471

*Articularis genus: a few separate muscle bundles arising deep to V. intermedius; tenses capsule of knee joint.

10. MEDIAL MUSCLES OF THE THIGH

Gracilis
Pectineus
Adductor longus
Adductor brevis
Adductor magnus

GRACILIS

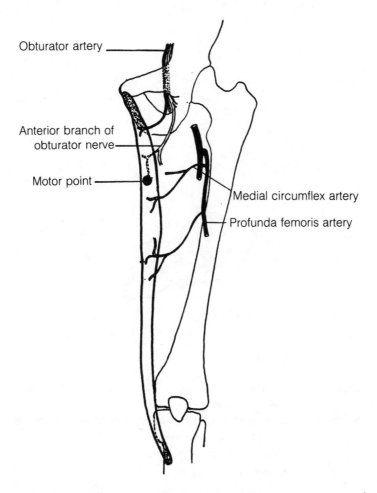

Obturator artery

Anterior branch of obturator nerve

Motor point

Medial circumflex artery

Profunda femoris artery

ORIGIN: Lower half of pubic symphysis, upper half of pubic arch
INSERTION: Upper part of medial surface of tibia
FUNCTION: Flexes leg and medially rotates at hip joint; adducts thigh
NERVE: Anterior branch of obturator
ARTERY: Muscular branches of profunda femoris, obturator, medial femoral circumflex

References

	GRAY	GRANT'S ATLAS	NETTER
Muscle	563	5-25	462, 465, 507
Nerve	563, 1226, 1230	5-24	467, 507
Artery	753, 767	5-24	474

PECTINEUS

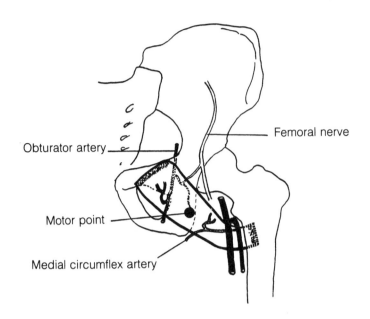

Femoral nerve

Obturator artery

Motor point

Medial circumflex artery

ORIGIN: Pectineal line, surface of pubis between iliopectineal eminence and pubic tubercle
INSERTION: Line extending from lesser trochanter to linea aspera
FUNCTION: Adducts, flexes, medially rotates thigh
NERVE: Muscular branches of femoral
ARTERY: Medial femoral circumflex, obturator

References

	GRAY	GRANT'S ATLAS	NETTER
Muscle	563	5-17	462, 470, 474
Nerve	563, 1226, 1232	5-24	467
Artery	753, 767	5-24	474, 481

ADDUCTOR LONGUS

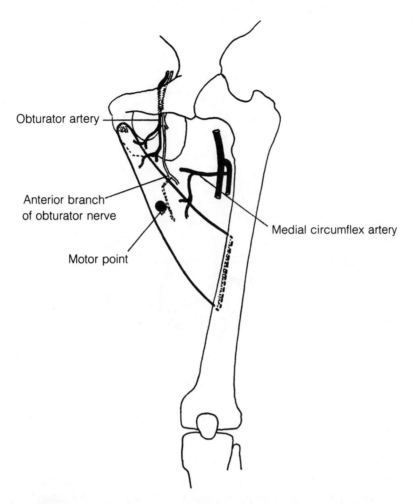

Obturator artery

Anterior branch
of obturator nerve

Motor point

Medial circumflex artery

ORIGIN: Front of pubis in angle between crest and symphysis
INSERTION: Middle half of medial lip of linea aspera
FUNCTION: Adducts thigh, and assists in flexing it. Rotator action is contro-
versial
NERVE: Anterior branch of obturator
ARTERY: Medial femoral circumflex, obturator

References

	GRAY	GRANT'S ATLAS	NETTER
Muscle	564	5-17	462, 470, 507
Nerve	564, 1226, 1230	5-24	467, 471, 507
Artery	753, 767	5-24	470

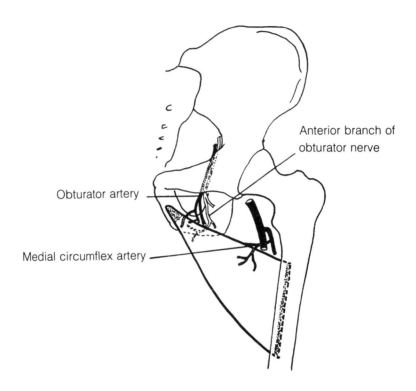

Anterior branch of
obturator nerve

Obturator artery

Medial circumflex artery

ORIGIN: Outer surface of inferior ramus of pubis
INSERTION: Line extending from lesser trochanter to linea aspera
FUNCTION: Adducts thigh, and assists in flexing it. Rotator action is contro-
             versial
NERVE: Anterior branch of obturator
ARTERY: Medial femoral circumflex, obturator

References

	GRAY	GRANT'S ATLAS	NETTER
Muscle	564	5-21	463, 471, 507
Nerve	564, 1226, 1230	Not shown	467, 471, 507
Artery	753, 767	5-24	471

ADDUCTOR MAGNUS

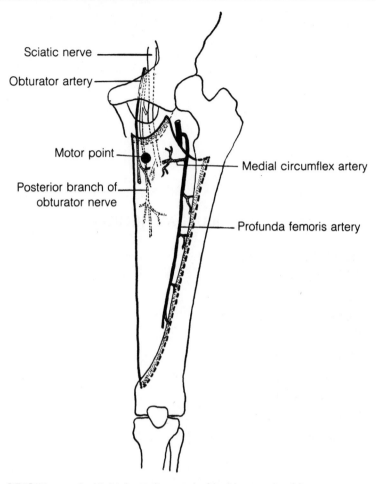

Sciatic nerve

Obturator artery

Motor point

Posterior branch of
obturator nerve

Medial circumflex artery

Profunda femoris artery

ORIGIN: Ischial tuberosity, rami of ischium and pubis
INSERTION: Line extending from greater trochanter to linea aspera, linea as-
pera, medial supracondylar line, adductor tubercle
FUNCTION: Adducts thigh; upper portion flexes it; lower portion extends it.
Rotating action is controversial
NERVE: Posterior branch of obturator, sciatic
ARTERY: Medial femoral circumflex, perforating branches of profunda fe-
moris, obturator, muscular branches of popliteal

References

	GRAY	GRANT'S ATLAS	NETTER
Muscle	565	5-34	463, 471, 507
Nerve	565, 1226, 1231, 1239	5-37	467, 471, 507
Artery	753, 767, 768	5-24, 5-37	471

11. MUSCLES OF THE GLUTEAL REGION

Gluteus maximus
Gluteus medius
Gluteus minimus
Tensor fasciae latae
Piriformis
Obturator internus
Gemellus superior
Gemellus inferior
Quadratus femoris
Obturator externus

GLUTEUS MAXIMUS

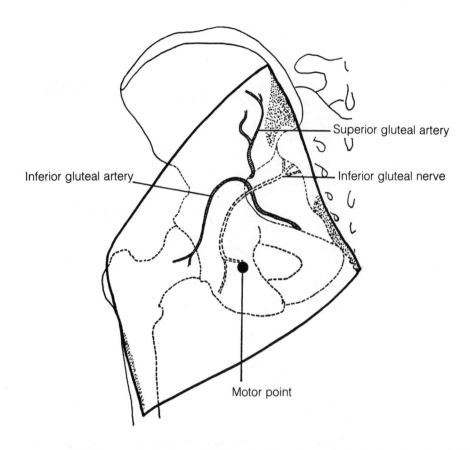

Superior gluteal artery

Inferior gluteal artery

Inferior gluteal nerve

Motor point

ORIGIN: Posterior gluteal line, tendon of sacrospinalis, dorsal surface of sacrum and coccyx, sacrotuberous ligament

INSERTION: Gluteal tuberosity of femur, iliotibial tract

FUNCTION: Extends thigh, assists in adducting and laterally rotating it; acting on insertion, muscle extends trunk

NERVE: Inferior gluteal

ARTERY: Superior gluteal, inferior gluteal, 1st perforating branch of profunda femoris

References

	GRAY	GRANT'S ATLAS	NETTER
Muscle	566	5-31	464, 465, 473
Nerve	567, 1235, 1236	5-36, 5-37	467, 473
Artery	757, 758, 768	5-36, 5-37	472

GLUTEUS MEDIUS

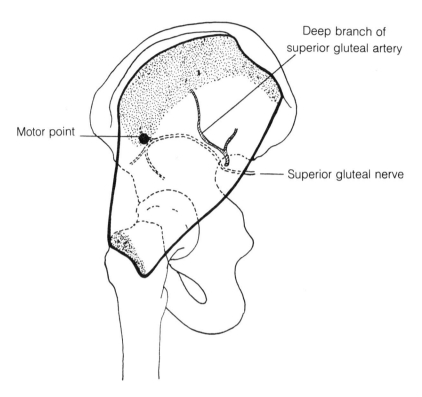

Deep branch of
superior gluteal artery

Motor point

Superior gluteal nerve

ORIGIN: Outer surface of ilium from iliac crest and posterior gluteal line above, to anterior gluteal line below, gluteal aponeurosis

INSERTION: Lateral surface of greater trochanter

FUNCTION: Abducts thigh, rotates thigh medially when limb is extended

NERVE: Superior gluteal

ARTERY: Deep branch of superior gluteal

References

	GRAY	GRANT'S ATLAS	NETTER
Muscle	567	5-33	465, 473
Nerve	568, 1235, 1236	5-37	467, 473
Artery	757	5-36	472

GLUTEUS MINIMUS

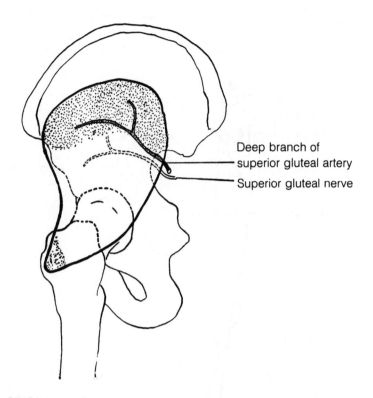

Deep branch of
superior gluteal artery

Superior gluteal nerve

ORIGIN: Outer surface of ilium between anterior and inferior gluteal lines margin of greater sciatic notch

INSERTION: Anterior border of greater trochanter

FUNCTION: Abducts thigh, rotates thigh medially when limb is extended

NERVE: Superior gluteal

ARTERY: Deep branch of superior gluteal

References

	GRAY	GRANT'S ATLAS	NETTER
Muscle	568	5-34, 5-37	465, 466, 473
Nerve	568, 1235, 1236	5-37	467, 473
Artery	757	5-37	472

TENSOR FASCIAE LATAE

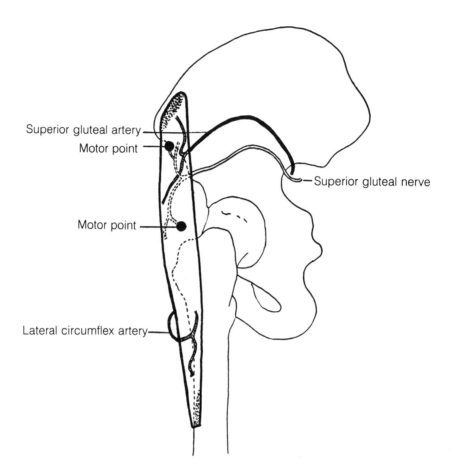

Superior gluteal artery

Motor point

Superior gluteal nerve

Motor point

Lateral circumflex artery

ORIGIN: Anterior part of outer lip of iliac crest, anterior border of ilium

INSERTION: Middle third of thigh along iliotibial tract

FUNCTION: Tenses fascia lata counteracting backward pull of gluteus maximus on iliotibial tract; assists in flexing, abducting, and medially rotating thigh

NERVE: Superior gluteal

ARTERY: Lateral femoral circumflex, superior gluteal

References

	GRAY	GRANT'S ATLAS	NETTER
Muscle	568	5-17	464, 470
Nerve	568, 1235, 1236	Not shown	467, 473
Artery	757, 767	5-17	472

PIRIFORMIS

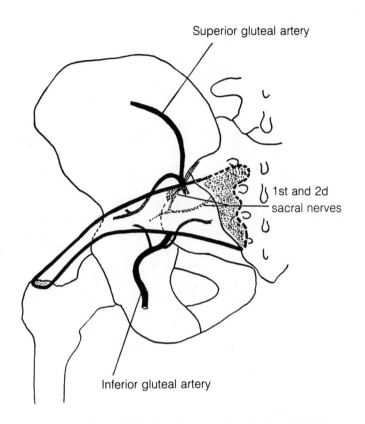

Superior gluteal artery

1st and 2d sacral nerves

Inferior gluteal artery

ORIGIN: Pelvic surface of sacrum between anterior sacral foramina, margin of greater sciatic foramen, sacrotuberous ligament
INSERTION: Upper border of greater trochanter of femur
FUNCTION: Rotates thigh laterally, abducts thigh when limb is flexed
NERVE: 1st and 2d sacral
ARTERY: Superior gluteal, inferior gluteal, internal pudendal

References

	GRAY	GRANT'S ATLAS	NETTER
Muscle	568	5-39	246, 339, 465, 473
Nerve	568, 1235, 1236	3-34	469
Artery	755, 757, 758	5-36, 5-37	472

OBTURATOR INTERNUS

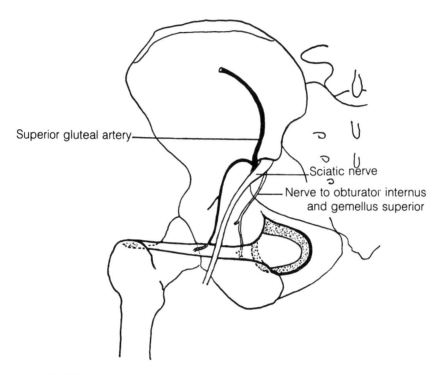

Superior gluteal artery

Sciatic nerve

Nerve to obturator internus and gemellus superior

ORIGIN: Margins of obturator foramen, obturator membrane, pelvic surface of hip bone behind and above obturator foramen, obturator fascia

INSERTION: Medial surface of greater trochanter

FUNCTION: Rotates thigh laterally, abducts thigh when limb is flexed

NERVE: Nerve to obturator internus and gemellus superior

ARTERY: Muscular branches of internal pudendal; superior gluteal

References

	GRAY	GRANT'S ATLAS	NETTER
Muscle	568	3-38, 5-41	337, 340, 465, 473
Nerve	570, 1235	5-36	467, 469, 473
Artery	755, 757	3-38	469

GEMELLUS SUPERIOR

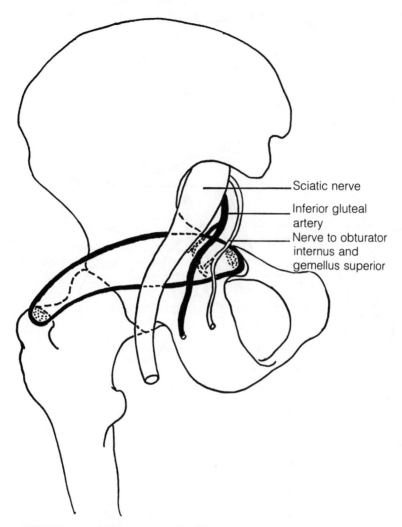

Sciatic nerve

Inferior gluteal artery

Nerve to obturator internus and gemellus superior

ORIGIN: Outer surface of ischial spine

INSERTION: Medial surface of greater trochanter, blends with obturator internus tendon

FUNCTION: Rotates thigh laterally

NERVE: Nerve to obturator internus and gemellus superior

ARTERY: Inferior gluteal

References

	GRAY	GRANT'S ATLAS	NETTER
Muscle	570	5-41	465, 472, 473
Nerve	570, 1235	5-36, 5-37	467, 469, 473
Artery	758	5-36, 5-37	377, 472

GEMELLUS INFERIOR

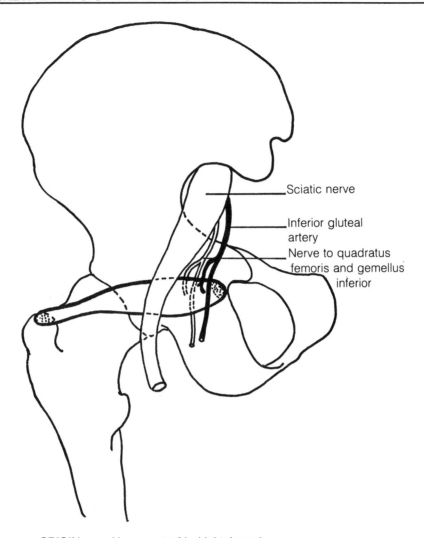

Sciatic nerve

Inferior gluteal artery

Nerve to quadratus femoris and gemellus inferior

ORIGIN: Upper part of ischial tuberosity
INSERTION: Medial surface of greater trochanter, blends with obturator internus tendon
FUNCTION: Rotates thigh laterally
NERVE: Nerve to quadratus femoris and gemellus inferior
ARTERY: Inferior gluteal

References

	GRAY	GRANT'S ATLAS	NETTER
Muscle	570	5-41	465, 472, 473
Nerve	570, 1235	Not shown	467, 469, 473
Artery	758	5-36, 5-37	472

QUADRATUS FEMORIS

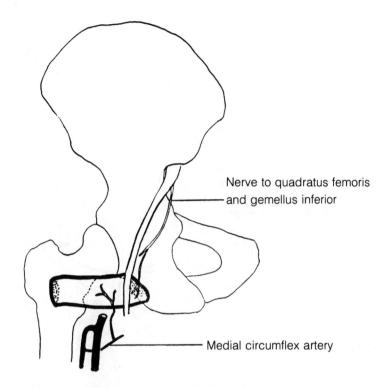

Nerve to quadratus femoris and gemellus inferior

Medial circumflex artery

ORIGIN: Lateral margin of ischial tuberosity
INSERTION: Quadrate tubercle of femur, linea quadrata
FUNCTION: Adducts and laterally rotates thigh
NERVE: Nerve to quadratus femoris and gemellus inferior
ARTERY: Medial femoral circumflex

References

	GRAY	GRANT'S ATLAS	NETTER
Muscle	570	5-39	463, 465, 473
Nerve	570, 1235	Not shown	467, 469, 473
Artery	767	5-36, 5-37	481

OBTURATOR EXTERNUS

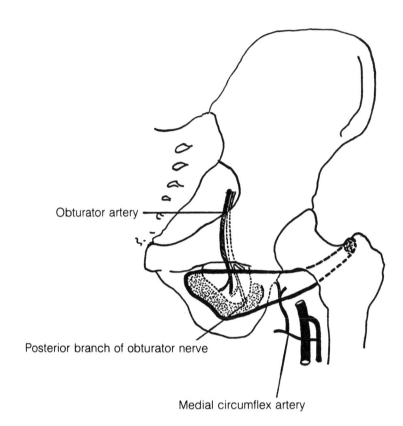

Obturator artery

Posterior branch of obturator nerve

Medial circumflex artery

ORIGIN: Outer margin of obturator foramen, outer surface of obturator membrane

INSERTION: Trochanteric fossa of femur

FUNCTION: Adducts thigh, rotates it laterally

NERVE: Posterior branch of obturator

ARTERY: Obturator, medial femoral circumflex

References

	GRAY	GRANT'S ATLAS	NETTER
Muscle	570	5-42	463, 507
Nerve	570, 1226, 1231	5-42	468, 507
Artery	753, 767	5-37	243

12. POSTERIOR MUSCLES OF THE THIGH

Biceps femoris
Semitendinosus
Semimembranosus

BICEPS FEMORIS

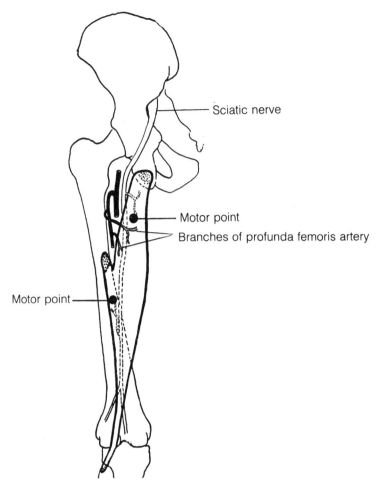

- Sciatic nerve
- Motor point
- Branches of profunda femoris artery

Motor point

ORIGIN: Long head from ischial tuberosity and sacrotuberous ligament; short head from lateral lip of linea aspera, lateral supracondylar line of femur, lateral intermuscular septum

INSERTION: Head of fibula, lateral condyle of tibia, deep fascia on lateral side of leg

FUNCTION: Flexes leg, extends thigh, rotates leg laterally when knee is semi-flexed

NERVE: Sciatic (tibial portion to long head, peroneal portion to short head)

ARTERY: Perforating branches of profunda femoris, superior muscular branches of popliteal

References

	GRAY	GRANT'S ATLAS	NETTER
Muscle	571	5-33, 5-34	464, 465
Nerve	572, 1235, 1239	5-36	467, 469, 472
Artery	768, 770	5-36, 5-37	472

SEMITENDINOSUS

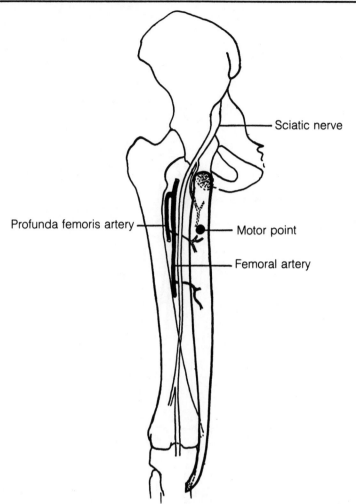

Sciatic nerve

Profunda femoris artery

Motor point

Femoral artery

ORIGIN: Upper and medial impression of ischial tuberosity with tendon of biceps

INSERTION: Upper part of medial surface of tibia, deep fascia of leg

FUNCTION: Flexes leg, extends thigh, rotates leg medially when knee is semiflexed

NERVE: Sciatic

ARTERY: Perforating branches of profunda femoris; superior muscular branches of popliteal

References

	GRAY	GRANT'S ATLAS	NETTER
Muscle	572	5-33	465, 472, 508
Nerve	572, 1235, 1239	5-36	467, 469, 472
Artery	768, 770	Not shown	Not shown

SEMIMEMBRANOSUS

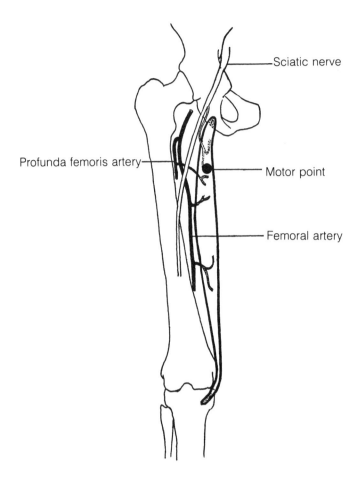

Sciatic nerve

Profunda femoris artery

Motor point

Femoral artery

ORIGIN: Upper and lateral facet of ischial tuberosity

INSERTION: Medial posterior surface of medial condyle of tibia

FUNCTION: Flexes leg, extends thigh, rotates leg medially when knee is semiflexed

NERVE: Sciatic

ARTERY: Perforating branches of profunda femoris; superior muscular branches of popliteal

References

	GRAY	GRANT'S ATLAS	NETTER
Muscle	572	5-33	465, 472, 508
Nerve	572, 1235, 1239	5-36	467, 469, 472
Artery	768, 770	5-37	Not shown

13. ANTERIOR MUSCLES OF THE LEG

Tibialis anterior
Extensor hallucis longus
Extensor digitorum longus
Peroneus tertius (Fibularis tertius)

TIBIALIS ANTERIOR

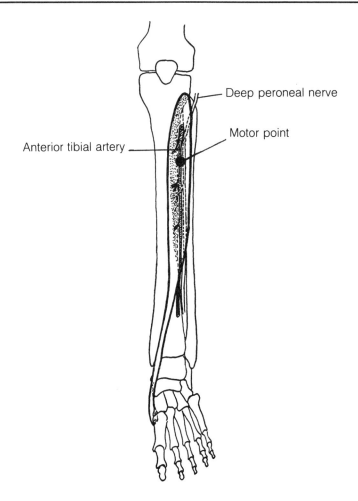

ORIGIN: Lateral condyle of tibia, upper two-thirds of lateral surface of tibia, interosseus membrane, deep fascia, lateral intermuscular septum

INSERTION: Medial and plantar surface of medial cuneiform bone, base of 1st metatarsal bone

FUNCTION: Dorsiflexes foot, inverts it

NERVE: Deep peroneal (anterior tibial)

ARTERY: Muscular branches of anterior tibial

References

	GRAY	GRANT'S ATLAS	NETTER
Muscle	573	5-78, 5-110	488, 499, 510
Nerve	574, 1241	5-78	489
Artery	774	5-78	489

EXTENSOR HALLUCIS LONGUS

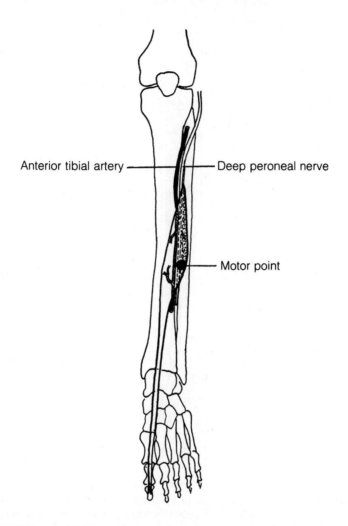

Anterior tibial artery — — Deep peroneal nerve

— Motor point

ORIGIN: Middle half of anterior surface of fibula, adjacent interosseous membrane

INSERTION: Base of distal phalanx of great toe

FUNCTION: Extends great toe, continued action dorsiflexes foot

NERVE: Deep peroneal (anterior tibial)

ARTERY: Muscular branches of anterior tibial

References

	GRAY	GRANT'S ATLAS	NETTER
Muscle	574	5-78, 5-82	488, 499, 510
Nerve	575, 1241	5-78	489
Artery	774	5-78	489

EXTENSOR DIGITORUM LONGUS

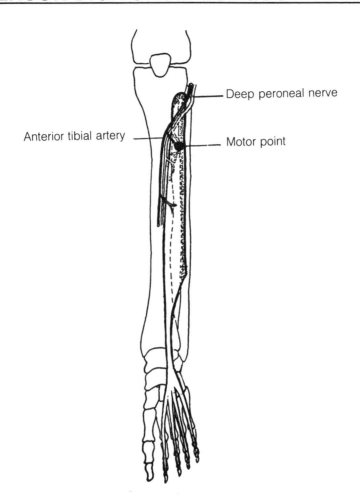

ORIGIN: Lateral condyle of tibia, upper three-fourths of anterior surface of fibula interosseous membrane, deep fascia, intermuscular septa

INSERTION: Dorsal surface of middle and distal phalanges of lateral 4 toes

FUNCTION: Extends phalanges of lateral 4 toes, continued action dorsiflexes foot

NERVE: Deep peroneal (anterior tibial)

ARTERY: Muscular branches of anterior tibial

References

	GRAY	GRANT'S ATLAS	NETTER
Muscle	575	5-78, 5-82	488, 490, 498
Nerve	575, 1241	5-78	489
Artery	774	5-78	489

PERONEUS TERTIUS (Fibularis tertius)

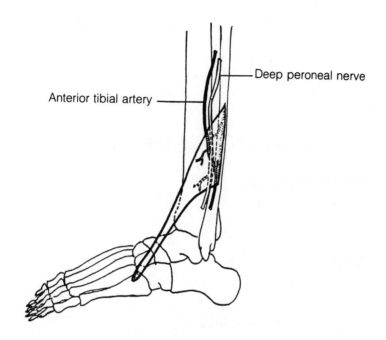

Anterior tibial artery

Deep peroneal nerve

ORIGIN: (Peroneus tertius is commonly known as the 5th tendon of extensor digitorum longus.) Lower anterior surface of fibula, adjacent intermuscular septum

INSERTION: Dorsal surface of base of 5th metatarsal bone

FUNCTION: Dorsiflexes and everts foot

NERVE: Deep peroneal (anterior tibial)

ARTERY: Muscular branches of anterior tibial

References

	GRAY	GRANT'S ATLAS	NETTER
Muscle	575	5-77, 5-78, 5-86, 5-88	490, 497, 498
Nerve	575, 1241	5-78	Not shown
Artery	774	Not shown	Not shown

14. POSTERIOR MUSCLES OF THE LEG

Superficial Group
Gastrocnemius
Soleus
Plantaris
Deep Group
Popliteus
Flexor hallucis longus
Flexor digitorum longus
Tibialis posterior

GASTROCNEMIUS

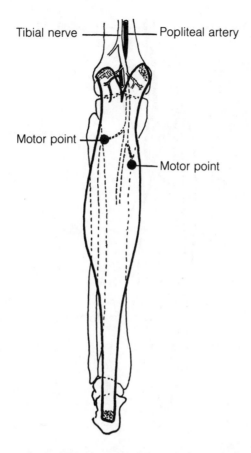

Tibial nerve ——— ——— Popliteal artery

Motor point ——— ———— Motor point

ORIGIN: <u>Medial head</u> from medial condyle and adjacent part of femur, capsule of knee joint; <u>lateral head</u> from lateral condyle and adjacent part of femur, capsule of knee joint

INSERTION: Into calcaneus by calcaneal tendon

FUNCTION: Plantarflexes foot; acting from below, flexes femur on tibia

NERVE: Tibial (medial popliteal)

ARTERY: Sural branches of popliteal

References

	GRAY	GRANT'S ATLAS	NETTER
Muscle	576	5-93	465, 485, 508
Nerve	576, 1239	5-54, 5-93	485, 508
Artery	770	5-55	Not shown

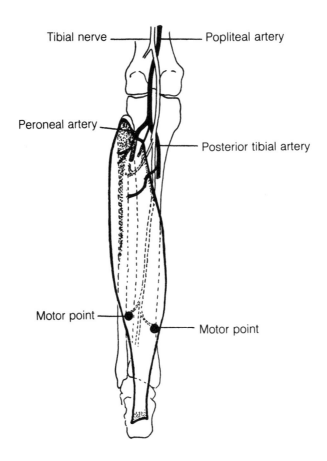

Tibial nerve — Popliteal artery

Peroneal artery — Posterior tibial artery

Motor point — Motor point

ORIGIN: Posterior surface of head and upper third of shaft of fibula, middle third of medial border of tibia, tendinous arch between tibia and fibula

INSERTION: Into calcaneus by calcaneal tendon

FUNCTION: Plantarflexes foot, steadies leg upon foot

NERVE: Tibial (medial popliteal)

ARTERY: Posterior tibial, peroneal, sural branches of popliteal

References

	GRAY	GRANT'S ATLAS	NETTER
Muscle	576	5-93, 5-94	486, 497
Nerve	577, 1239	5-54, 5-94	508
Artery	770, 777, 779	5-94	487

PLANTARIS

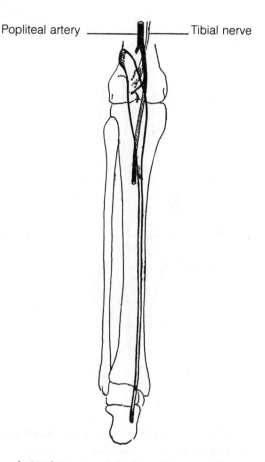

Popliteal artery —————— Tibial nerve

ORIGIN:	Lateral supracondylar line of femur, oblique popliteal ligament of knee joint
INSERTION:	Medial side of posterior part of calcaneus, calcaneal tendon
FUNCTION:	Plantarflexes foot
NERVE:	Tibial (medial popliteal)
ARTERY:	Sural branches of popliteal

References

	GRAY	GRANT'S ATLAS	NETTER
Muscle	577	5-54	465, 485, 486
Nerve	577, 1239	Not shown	486
Artery	770	Not shown	486

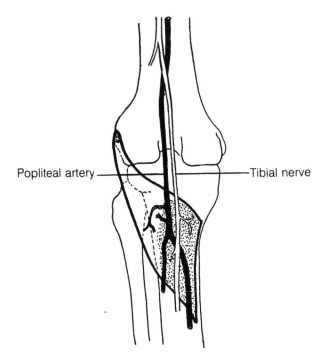

Popliteal artery ———————————— Tibial nerve

ORIGIN: Lateral condyle of femur, oblique popliteal ligament of knee
INSERTION: Triangular area on posterior surface of tibia above soleal line
FUNCTION: Flexes leg, rotates tibia medially at beginning of flexion
NERVE: Tibial (medial or internal popliteal)
ARTERY: Genicular branches of popliteal

References

	GRAY	GRANT'S ATLAS	NETTER
Muscle	577	5-55, 5-68, 5-96	465, 486, 487
Nerve	577, 1239	5-54, 5-96	Not shown
Artery	772	5-55	486

FLEXOR HALLUCIS LONGUS

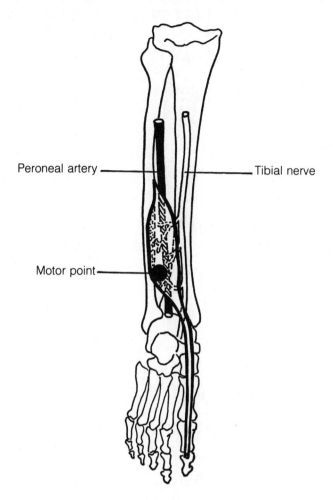

Peroneal artery —

— Tibial nerve

Motor point —

ORIGIN: Lower two thirds of posterior surface of fibula, interosseous membrane, adjacent intermuscular septa and fascia

INSERTION: Base of distal phalanx of great toe

FUNCTION: Flexes great toe, continued action aids in plantarflexing foot

NERVE: Tibial (medial or internal popliteal)

ARTERY: Muscular branches of peroneal

References

	GRAY	GRANT'S ATLAS	NETTER
Muscle	578	5-95, 5-102	487, 502, 509
Nerve	578, 1239	5-96	487, 509
Artery	777	5-95	487

FLEXOR DIGITORUM LONGUS

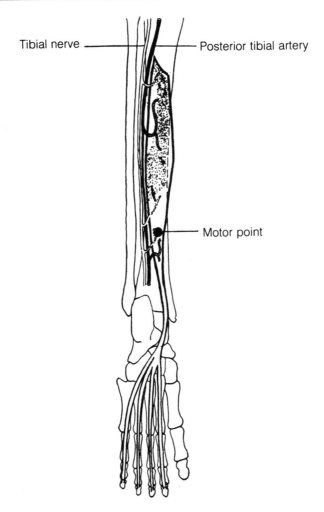

Tibial nerve — — Posterior tibial artery

— Motor point

ORIGIN: Posterior surface of middle three fifths of tibia, fascia covering tibialis posterior

INSERTION: Plantar surface of base of distal phalanx of lateral 4 toes

FUNCTION: Flexes phalanges of lateral 4 toes, continued action plantarflexes foot

NERVE: Tibial (medial or internal popliteal)

ARTERY: Posterior tibial

References

	GRAY	GRANT'S ATLAS	NETTER
Muscle	578	5-96, 5-107	487, 502, 509
Nerve	579, 1239	5-96	487, 509
Artery	779	5-95	487

TIBIALIS POSTERIOR

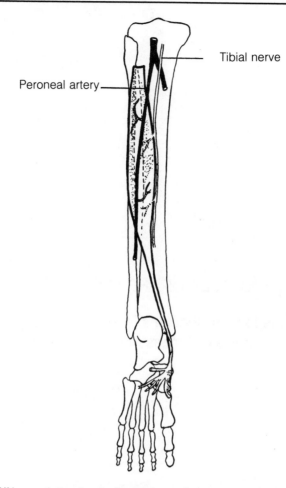

Tibial nerve

Peroneal artery

ORIGIN: Lateral part of posterior surface of tibia, upper two thirds of medial surface of fibula, deep transverse fascia, adjacent intermuscular septa, posterior surface of interosseus membrane

INSERTION: Tuberosity of navicular bone, plantar surface of all cuneiform bones, plantar surface of base of 2d, 3d, and 4th metatarsal bones, cuboid bone, sustentaculum tali

FUNCTION: Plantarflexes foot, inverts it

NERVE: Tibial (medial or internal popliteal)

ARTERY: Peroneal

References

	GRAY	GRANT'S ATLAS	NETTER
Muscle	579	5-96, 5-110	487, 495, 502
Nerve	579, 1239	5-96	487, 509
Artery	777	5-96	487

15. LATERAL MUSCLES OF THE LEG

Peroneus longus (Fibularis longus)
Peroneus brevis (Fibularis brevis)

PERONEUS LONGUS (Fibularis longus)

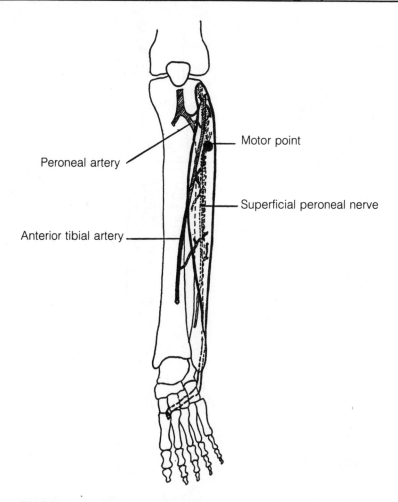

Peroneal artery

Motor point

Superficial peroneal nerve

Anterior tibial artery

ORIGIN: Lateral condyle of tibia, head and upper two thirds of lateral surface of fibula, adjacent fascia, intermuscular septa

INSERTION: Lateral side of medial cuneiform bone, base of 1st metatarsal bone

FUNCTION: Plantarflexes foot, everts it

NERVE: Superficial peroneal (musculocutaneous)

ARTERY: Muscular branches of anterior tibial, muscular branches of peroneal

References

	GRAY	GRANT'S ATLAS	NETTER
Muscle	579	5-77, 5-88, 5-128	490, 503, 510
Nerve	580, 1243	5-80	489, 510
Artery	777	Not shown	Not shown

PERONEUS BREVIS (Fibularis brevis)

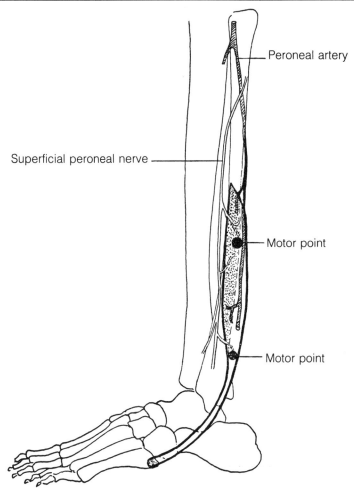

Peroneal artery

Superficial peroneal nerve

Motor point

Motor point

ORIGIN: Lower two-thirds of lateral surface of fibula, adjacent intermuscular septa

INSERTION: Lateral side of base of 5th metatarsal bone

FUNCTION: Plantarflexes foot, everts it

NERVE: Superficial peroneal (musculocutaneous)

ARTERY: Muscular branches of peroneal

References

	GRAY	GRANT'S ATLAS	NETTER
Muscle	580	5-77, 5-88	490, 503, 510
Nerve	580, 1243	5-80	489, 510
Artery	777	Not shown	Not shown

16. MUSCLES OF THE FOOT

Extensor digitorum brevis
Abductor hallucis
Flexor digitorum brevis
Abductor digiti minimi
Quadratus plantae (Flexor Accessorius)
Lumbricales
Flexor hallucis brevis
Adductor hallucis
Flexor digiti minimi brevis
Interossei dorsales (Dorsal Interossei)
Interossei plantares (Plantar Interossei)

EXTENSOR DIGITORUM BREVIS

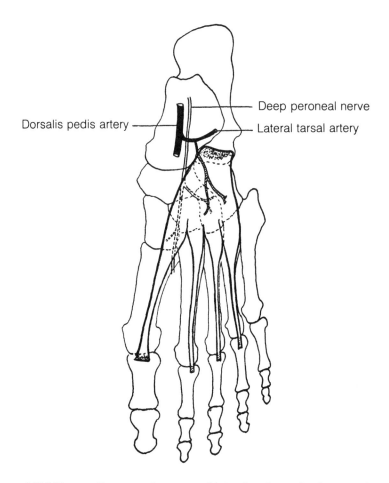

Dorsalis pedis artery

Deep peroneal nerve

Lateral tarsal artery

ORIGIN: Forepart of upper and lateral surface of calcaneus, lateral talo-calcaneal ligament, cruciate crural ligament

INSERTION: 1st tendon into dorsal surface of base of proximal phalanx of great toe, remaining 3 tendons into lateral sides of tendons of extensor digitorum longus

FUNCTION: Extends phalanges of 4 medial toes

NERVE: Deep peroneal (anterior tibial)

ARTERY: Dorsalis pedis, lateral tarsal

References

	GRAY	GRANT'S ATLAS	NETTER
Muscle	584	5-82, 5-86	498, 510
Nerve	584, 1241	5-78	499, 510
Artery	775	Not shown	499

ABDUCTOR HALLUCIS

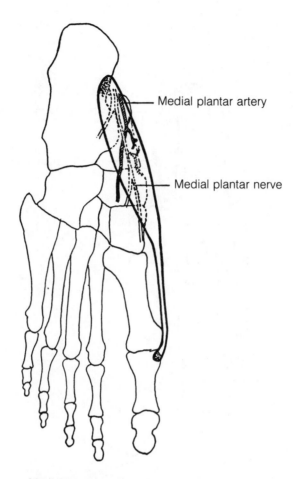

Medial plantar artery

Medial plantar nerve

ORIGIN: Medial process of calcaneus, laciniate ligament, plantar aponeurosis, adjacent intermuscular septum

INSERTION: Medial side of base of proximal phalanx of great toe

FUNCTION: Abducts great toe

NERVE: Medial plantar

ARTERY: Medial plantar

References

	GRAY	GRANT'S ATLAS	NETTER
Muscle	586	5-113	501, 509
Nerve	586, 1240	5-100, 5-112	501, 509
Artery	779	5-100	503

FLEXOR DIGITORUM BREVIS

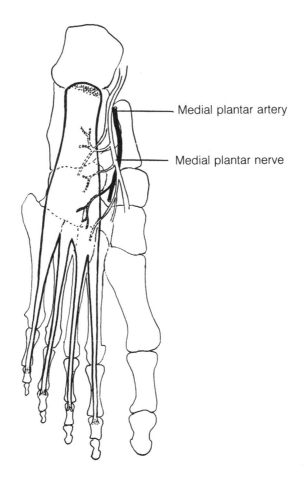

Medial plantar artery

Medial plantar nerve

ORIGIN: Medial process of calcaneus, plantar aponeurosis, adjacent intermuscular septa

INSERTION: Middle phalanx of lateral 4 toes

FUNCTION: Flexes middle phalanges on proximal, continued action also flexes proximal phalanges of lateral 4 toes

NERVE: Medial plantar

ARTERY: Medial plantar

References

	GRAY	GRANT'S ATLAS	NETTER
Muscle	586	5-105	501, 509
Nerve	586, 1240	Not shown	509
Artery	779	Not shown	Not shown

ABDUCTOR DIGITI MINIMI

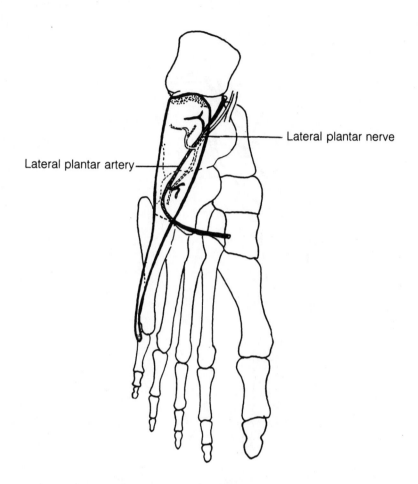

Lateral plantar nerve

Lateral plantar artery

ORIGIN: Lateral and medial processes of calcaneus, calcaneal fascia, adjacent intermuscular septum

INSERTION: Lateral side of base of proximal phalanx of little toe

FUNCTION: Abducts little toe, assists in flexing it

NERVE: Lateral plantar

ARTERY: Lateral plantar

References

	GRAY	GRANT'S ATLAS	NETTER
Muscle	586	5-113	499, 501, 509
Nerve	587, 1241	Not shown	509
Artery	779	Not shown	Not shown

QUADRATUS PLANTAE (Flexor Accessorius)

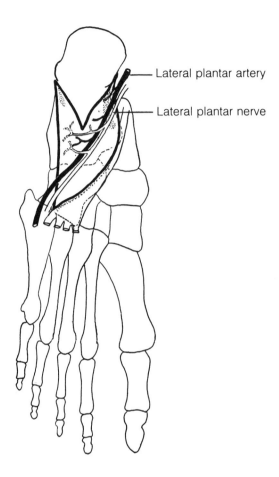

— Lateral plantar artery

— Lateral plantar nerve

ORIGIN: <u>Medial head</u> from medial surface of calcaneus and medial border of long plantar ligament, <u>lateral head</u> from lateral border of plantar surface of calcaneus and lateral border of long plantar ligament

INSERTION: Tendons of flexor digitorum longus

FUNCTION: Flexes terminal phalanges of lateral 4 toes

NERVE: Lateral plantar

ARTERY: Lateral plantar

References

	GRAY	GRANT'S ATLAS	NETTER
Muscle	587	5-107, 5-110	502, 509
Nerve	587, 1241	5-112	502, 509
Artery	779	5-112	503

LUMBRICALES

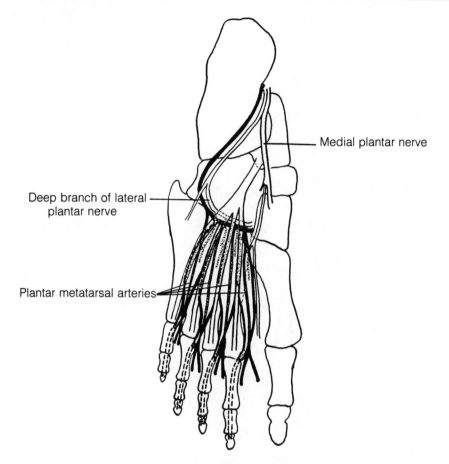

Medial plantar nerve

Deep branch of lateral plantar nerve

Plantar metatarsal arteries

ORIGIN: There are 4 lumbricales, all arising from tendons of flexor digitorum longus: 1st from medial side of tendon for 2d toe, 2d from adjacent sides of tendons for 2d and 3d toes, 3d from adjacent sides of tendons for 3d and 4th toes, 4th from adjacent sides of tendons for 4th and 5th toes

INSERTION: With tendons of extensor digitorum longus and interossei into bases of terminal phalanges of 4 lateral toes

FUNCTION: Flex toes at metatarsophalangeal joints, extend toes at interphalangeal joints

NERVE: Medial plantar, deep lateral plantar

ARTERY: Plantar metatarsal

References

	GRAY	GRANT'S ATLAS	NETTER
Muscle	588	5-107	501, 509
Nerve	588, 1240, 1241	Not shown	509
Artery	780	5-105	503

FLEXOR HALLUCIS BREVIS

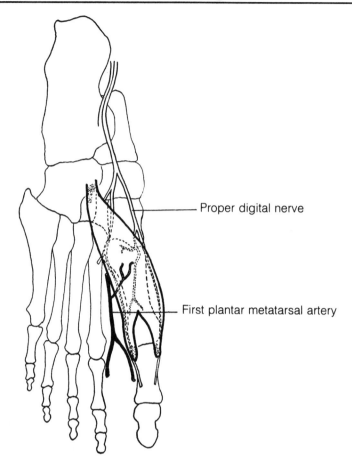

Proper digital nerve

First plantar metatarsal artery

ORIGIN: Medial part of plantar surface of cuboid bone, adjacent portion of lateral cuneiform bone, prolongation of tendon of tibialis posterior

INSERTION: Medial and lateral side of proximal phalanx of great toe

FUNCTION: Flexes great toe

NERVE: Proper digital nerve of great toe (1st plantar digital nerve) of medial plantar nerve

ARTERY: First plantar metatarsal (from junction of lateral and deep plantar arteries)

References

	GRAY	GRANT'S ATLAS	NETTER
Muscle	588	5-113	501, 503, 509
Nerve	588, 1240	5-112	509
Artery	780	Not shown	503

ADDUCTOR HALLUCIS

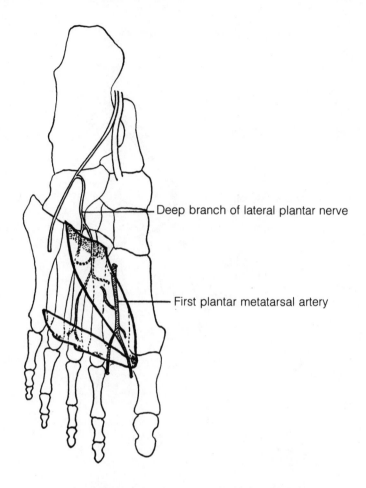

Deep branch of lateral plantar nerve

First plantar metatarsal artery

ORIGIN: Oblique head from bases of 2d, 3d, and 4th metatarsal bones, sheath of peroneus longus; transverse head from capsules of 2d, 3d, 4th, and 5th metatarsophalangeal ligaments, transverse ligament of sole

INSERTION: Lateral side of base of proximal phalanx of great toe

FUNCTION: Adducts great toe, assists in flexing it

NERVE: Deep branch of lateral plantar

ARTERY: First plantar metatarsal

References

	GRAY	GRANT'S ATLAS	NETTER
Muscle	589	5-112	503
Nerve	589, 1241	5-112	503
Artery	780	Not shown	503

FLEXOR DIGITI MINIMI BREVIS

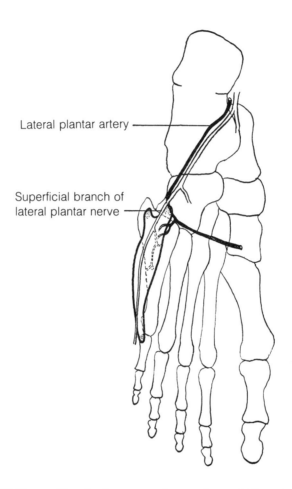

Lateral plantar artery

Superficial branch of
lateral plantar nerve

ORIGIN: Sheath of peroneus longus, base of 5th metatarsal bone
INSERTION: Lateral side of base of proximal phalanx of little toe
FUNCTION: Flexes little toe
NERVE: Superficial branch of lateral plantar
ARTERY: Lateral plantar

References

	GRAY	GRANT'S ATLAS	NETTER
Muscle	589	5-113	501, 502, 509
Nerve	589, 1241	5-112	502, 503
Artery	779	Not shown	503

INTEROSSEI DORSALES (Dorsal Interossei)

Superficial branch of lateral plantar nerve

Deep branch of lateral plantar nerve

Dorsal metatarsal arteries

ORIGIN: There are 4 dorsal interossei, each arising by 2 heads from adjacent sides of metatarsal bones

INSERTION: 1st into medial side of proximal phalanx of 2d toe, 2d into lateral side of proximal phalanx of 2d toe, 3d into lateral side of proximal phalanx of 3d toe, 4th into lateral side of proximal phalanx of 4th toe

FUNCTION: Abduct 2d, 3d, and 4th toes from axis of 2d toe, assist in flexing proximal phalanges and in extending middle and distal phalanges

NERVE: Superficial and deep branches of lateral plantar

ARTERY: Dorsal metatarsal

References

	GRAY	GRANT'S ATLAS	NETTER
Muscle	589	5-82, 5-113	499, 504, 505
Nerve	590, 1241	5-112	Not shown
Artery	776	5-80	499

INTEROSSEI PLANTARES (Plantar Interossei)

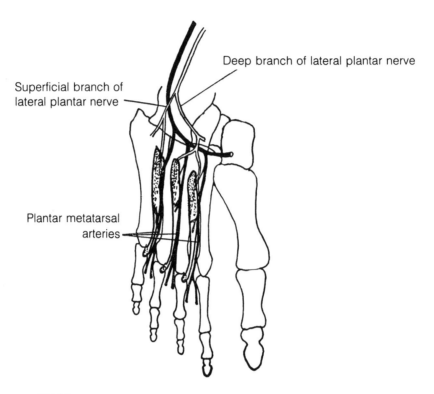

Deep branch of lateral plantar nerve

Superficial branch of
lateral plantar nerve

Plantar metatarsal
arteries

ORIGIN: There are 3 plantar interossei, arising from bases and medial sides of 3d, 4th, and 5th metatarsal bones

INSERTION: Medial sides of bases of proximal phalanges of 3d, 4th, and 5th toes

FUNCTION: Adduct 3d, 4th, and 5th toes toward axis of 2d toe, assist in flexing proximal phalanges and in extending middle and distal phalanges

NERVE: Superficial and deep branches of lateral plantar

ARTERY: Plantar metatarsal

References

	GRAY	GRANT'S ATLAS	NETTER
Muscle	590	5-113	503, 505
Nerve	590, 1241	5-112	Not shown
Artery	780	Not shown	503

CHARTS

Scheme of the Brachial Plexus
Nerves of upper extremities
Arteries of upper extremities
Nerves of lower extremities
Arteries of lower extremities
Muscles of right upper extremity—anterior view
Muscles of right upper extremity—posterior view
Muscles of right lower extremity—anterior view
Muscles of right lower extremity—posterior view

SCHEME OF THE BRACHIAL PLEXUS

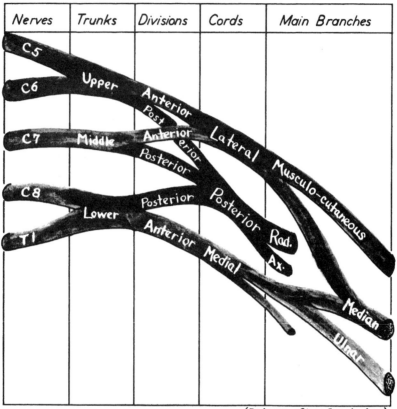

Nerves	Trunks	Divisions	Cords	Main Branches

(Redrawn after Cunningham)

References

GRAY	GRANT'S ATLAS	NETTER
1205, 1214	6-27, 8-4, 8-7, 8-51	148, 168, 173, 405, 408

NERVES OF UPPER EXTREMITIES

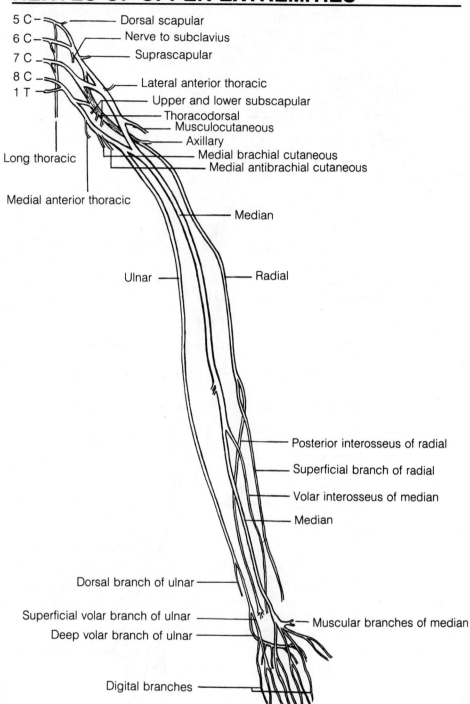

5 C — Dorsal scapular
6 C — Nerve to subclavius
7 C — Suprascapular
8 C — Lateral anterior thoracic
1 T — Upper and lower subscapular
Thoracodorsal
Musculocutaneous
Axillary
Medial brachial cutaneous
Medial antibrachial cutaneous

Long thoracic

Medial anterior thoracic

Median

Ulnar — Radial

Posterior interosseus of radial

Superficial branch of radial

Volar interosseus of median

Median

Dorsal branch of ulnar

Superficial volar branch of ulnar
Deep volar branch of ulnar — Muscular branches of median

Digital branches

129

ARTERIES OF UPPER EXTREMITIES

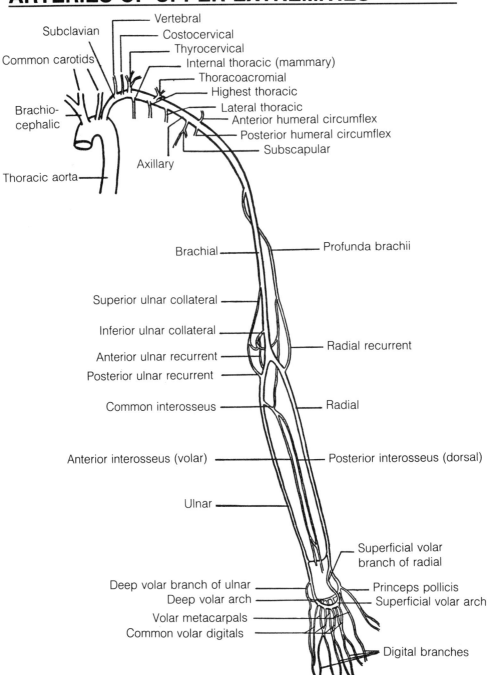

Subclavian

Common carotids

Brachio-
cephalic

Vertebral

Costocervical

Thyrocervical

Internal thoracic (mammary)

Thoracoacromial

Highest thoracic

Lateral thoracic

Anterior humeral circumflex

Posterior humeral circumflex

Subscapular

Axillary

Thoracic aorta

Brachial

Profunda brachii

Superior ulnar collateral

Inferior ulnar collateral

Anterior ulnar recurrent

Posterior ulnar recurrent

Common interosseus

Radial recurrent

Radial

Anterior interosseus (volar)

Posterior interosseus (dorsal)

Ulnar

Superficial volar
branch of radial

Deep volar branch of ulnar

Deep volar arch

Volar metacarpals

Common volar digitals

Princeps pollicis

Superficial volar arch

Digital branches

NERVES OF LOWER EXTREMITIES

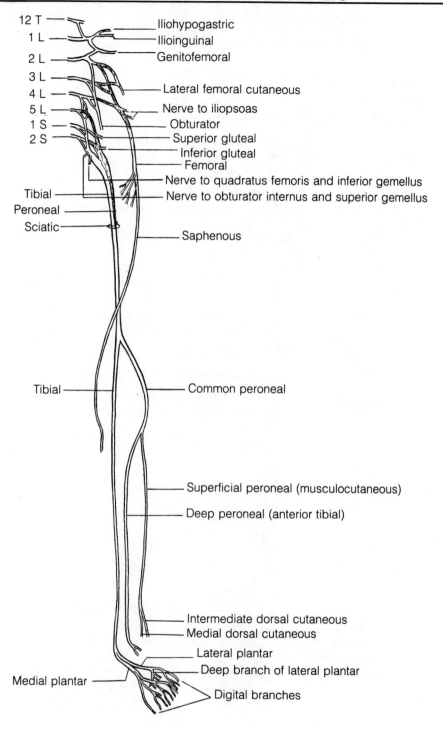

12 T — Iliohypogastric
1 L — Ilioinguinal
2 L — Genitofemoral
3 L
4 L — Lateral femoral cutaneous
5 L — Nerve to iliopsoas
1 S — Obturator
2 S — Superior gluteal
— Inferior gluteal
— Femoral
— Nerve to quadratus femoris and inferior gemellus
Tibial — — Nerve to obturator internus and superior gemellus
Peroneal —
Sciatic — — Saphenous

Tibial — — Common peroneal

— Superficial peroneal (musculocutaneous)

— Deep peroneal (anterior tibial)

— Intermediate dorsal cutaneous
— Medial dorsal cutaneous
— Lateral plantar
— Deep branch of lateral plantar
Medial plantar — — Digital branches

ARTERIES OF LOWER EXTREMITIES

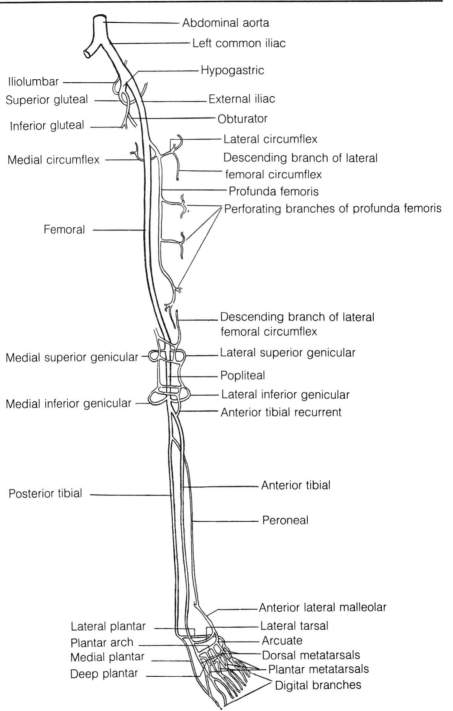

Abdominal aorta

Left common iliac

Hypogastric

Iliolumbar

Superior gluteal

External iliac

Inferior gluteal

Obturator

Lateral circumflex

Medial circumflex

Descending branch of lateral
femoral circumflex

Profunda femoris

Perforating branches of profunda femoris

Femoral

Descending branch of lateral
femoral circumflex

Medial superior genicular

Lateral superior genicular

Popliteal

Lateral inferior genicular

Medial inferior genicular

Anterior tibial recurrent

Posterior tibial

Anterior tibial

Peroneal

Anterior lateral malleolar

Lateral plantar

Lateral tarsal

Plantar arch

Arcuate

Medial plantar

Dorsal metatarsals

Deep plantar

Plantar metatarsals

Digital branches

MUSCLES OF RIGHT UPPER EXTREMITY— ANTERIOR VIEW

ORIGINS—SOLID   INSERTIONS—STIPPLED

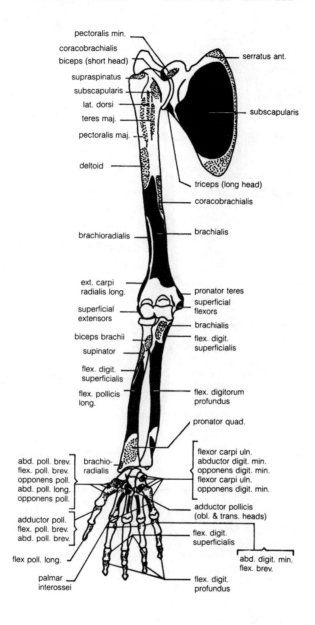

pectoralis min.

coracobrachialis

biceps (short head)

supraspinatus

subscapularis

lat. dorsi

teres maj.

pectoralis maj.

deltoid

brachioradialis

ext. carpi
radialis long.

superficial
extensors

biceps brachii

supinator

flex. digit.
superficialis

flex. pollicis
long.

abd. poll. brev.
flex. poll. brev.
opponens poll.
abd. poll. long.
opponens poll.

brachio-
radialis

adductor poll.
flex. poll. brev.
abd. poll. brev.

flex poll. long.

palmar
interossei

serratus ant.

subscapularis

triceps (long head)

coracobrachialis

brachialis

pronator teres
superficial
flexors

brachialis

flex. digit.
superficialis

flex. digitorum
profundus

pronator quad.

flexor carpi uln.
abductor digit. min.
opponens digit. min.
flexor carpi uln.
opponens digit. min.

adductor pollicis
(obl. & trans. heads)

flex. digit.
superficialis

abd. digit. min.
flex. brev.

flex. digit.
profundus

MUSCLES OF RIGHT UPPER EXTREMITY— POSTERIOR VIEW

ORIGINS—SOLID   INSERTIONS—STIPPLED

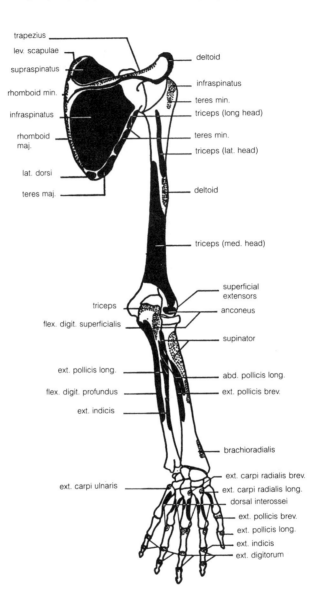

trapezius

lev. scapulae

supraspinatus

rhomboid min.

infraspinatus

rhomboid maj.

lat. dorsi

teres maj.

deltoid

infraspinatus

teres min.

triceps (long head)

teres min.

triceps (lat. head)

deltoid

triceps (med. head)

triceps

flex. digit. superficialis

ext. pollicis long.

flex. digit. profundus

ext. indicis

ext. carpi ulnaris

superficial extensors

anconeus

supinator

abd. pollicis long.

ext. pollicis brev.

brachioradialis

ext. carpi radialis brev.

ext. carpi radialis long.

dorsal interossei

ext. pollicis brev.

ext. pollicis long.

ext. indicis

ext. digitorum

MUSCLES OF RIGHT LOWER EXTREMITY—
ANTERIOR VIEW

ORIGINS—SOLID   INSERTIONS—STIPPLED

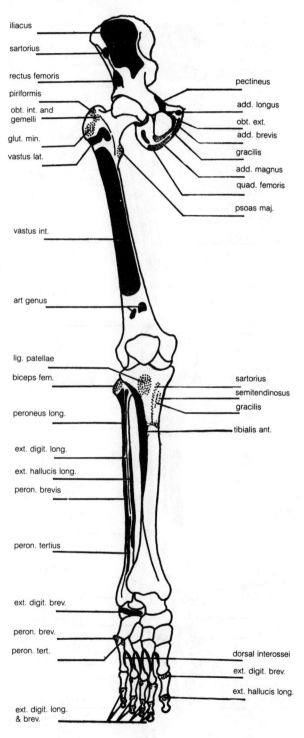

iliacus

sartorius

rectus femoris

piriformis

obt. int. and
gemelli

glut. min.

vastus lat.

vastus int.

art genus

lig. patellae

biceps fem.

peroneus long.

ext. digit. long.

ext. hallucis long.

peron. brevis

peron. tertius

ext. digit. brev.

peron. brev.

peron. tert.

ext. digit. long.
& brev.

pectineus

add. longus

obt. ext.

add. brevis

gracilis

add. magnus

quad. femoris

psoas maj.

sartorius

semitendinosus

gracilis

tibialis ant.

dorsal interossei

ext. digit. brev.

ext. hallucis long.

MUSCLES OF RIGHT LOWER EXTREMITY—
POSTERIOR VIEW

ORIGINS—SOLID   INSERTIONS—STIPPLED

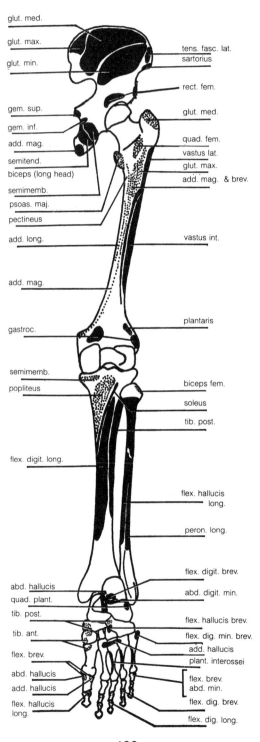

glut. med.

glut. max.

glut. min.

gem. sup.

gem. inf.

add. mag.

semitend.

biceps (long head)

semimemb.

psoas. maj.

pectineus

add. long.

add. mag.

gastroc.

semimemb.

popliteus

flex. digit. long.

abd. hallucis

quad. plant.

tib. post.

tib. ant.

flex. brev.

abd. hallucis

add. hallucis

flex. hallucis
long.

tens. fasc. lat.

sartorius

rect. fem.

glut. med.

quad. fem.

vastus lat.

glut. max.

add. mag. & brev.

vastus int.

plantaris

biceps fem.

soleus

tib. post.

flex. hallucis
long.

peron. long.

flex. digit. brev.

abd. digit. min.

flex. hallucis brev.

flex. dig. min. brev.

add. hallucis

plant. interossei

flex. brev.
abd. min.

flex. dig. brev.

flex. dig. long.

INDEX

INDEX

Rectus femoris, 74
Rhomboid major, 14
Rhomboid minor, 15
Sartorius, 73
Semimembranosus, 98
Semitendinosus, 97
Serratus anterior, 21
Soleus, 106
Subclavius, 20
Subscapularis, 24
Supinator, 51
Supraspinatus, 25
Tensor fasciae latae, 88
Teres major, 28
Teres minor, 27
Tibialis anterior, 100
Tibialis posterior, 111
Trapezius, 12
Triceps brachii, 33
Vastus intermedius, 77
Vastus lateralis, 75
Vastus medialis, 76
Volar interossei (hand), 67

NERVES

Anterior interosseus: see Median:
    Volar i.
Anterior tibial: see Deep peroneal
Axillary:
    Anterior branch, 23
    Posterior branch, 23, 27
Cervical:
    3d, 12, 16
    4th, 12, 16
    5th, 20
    6th, 20
Deep peroneal, 100-103, 116
Dorsal interosseus: see Radial:
    Posterior i.
Dorsal scapular, 14-16
Femoral, 71-77, 80
First plantar digital: see Proper
    digital
Gluteal: see Inferior g., Superior g.
Inferior gluteal, 85
Lateral anterior thoracic, 18

Lateral plantar, 119, 120
    Deep branch, 121, 123, 125, 126
    Superficial branch, 124-126
Long thoracic, 21
Lumbar:
    1st, 70
    2d, 69, 70
    3d, 69
Medial anterior thoracic, 18, 19
Medial plantar, 117, 118, 121
Median, 35-37, 39, 57-59, 65
Median, Volar interosseus, 40-42
Musculocutaneous:
    Arm, 30–32
    Leg: see Superficial peroneal
Nerve charts: see under CHARTS
Nerve to obturator internus and
        gemellus superior, 90, 91
Nerve to quadratus femoris and
        gemellus inferior, 92, 93
Obturator:
    Anterior branch, 79, 81, 82
    Posterior branch, 83, 94
Peroneal: see Deep p., Superficial p.
Plantar: see Lateral p., Medial p.
Posterior tibial: see Tibial
Proper digital nerve of great toe, 122
Radial, 32, 33, 44, 45, 49–51
    Posterior interosseus, 46-49, 52-55
Sacral:
    1st, 89
    2d, 89
Scapular: see Dorsal s.
Sciatic, 83, 90-92, 96-98
Spinal accessory, 12
Subscapular:
    Lower, 24, 28
    Upper, 24
Superficial peroneal, 113, 114
Superior gluteal, 86-88
Suprascapular, 25, 26
Thoracic: see Lateral anterior t.,
        Long t., Medial anterior t.
Thoracodorsal, 13
Tibial, 105-111
Ulnar, 38, 40
    Deep volar branch, 59, 60, 62-67
    Superficial volar branch, 61